Level C Science

Measuring Up®

to the

New York State Learning Standards

and Success Strategies for the State Test

This book is customized for New York and the lessons match the New York State Learning Standards. The Measuring Up® program includes instructional worktexts, Measuring Up e-Path® online formative assessment, and print Diagnostic Practice Tests, which are available separately.

800-822-1080
www.PeoplesEducation.com

Peoples Education™
Your partner in student success™

Executive Vice President, Chief Creative Officer: Diane Miller
Vice President, Product Development: Steven Jay Griffel
Assistant Vice President, Editorial Director: Eugene McCormick
Editorial Development: The Publisher's Partnership
Executive Editor: Michael Goodman
Senior Editor: Scott Caffrey
Editor: Kerri Gero
Supervising Copy Editor: Lee Laddy
Copy Editors: Katy Leclercq, Basanti Das, Joe Schwab, Josh Gillenson, Kathy McCormick
Vice President of Marketing: Victoria Ameer Kiely
Director of Marketing: Melissa Dubno Geller
Vice President, Production and Manufacturing: Doreen Smith
Production Manager, Science & Mathematics: Jennifer Brewer
Assistant Production Manager, Science & Mathematics: Jennifer Tully
Project Manager, Science & Mathematics: Kenneth Owens
Production Editors: Sharon MacGregor, Eileen Turano, Swavek Zielinski
Art Manager: Salvatore Esposito
Permissions Manager: Kristine Liebman
Cover Design: Yadiro Henriquez, Cynthia Mackowicz, Michele Sakow, Chris Kennedy

New York Advisory Panel:

Margaret Masi, Reading Specialist, School 17, Yonkers Public Schools

Carrie A. Sunkes, Coordinator of Gifted and Talented Programs, Burnt Hills, Ballston Lake School District

Cily Rueda, LEP/Math Specialist, Blue Creek Elementary, Latham

Judy Fix, Bennett Park Montessori #2, Buffalo

Peoples Education™
Your partner in student success™

Copyright © 2008
Peoples Education, Inc.
299 Market Street
Saddle Brook, New Jersey 07663

ISBN 978-1-4138-8126-4

All rights reserved. No part of this book may be kept in an information storage or retrieval system, transmitted or reproduced in any form or by any means without prior written permission of the Publisher.

Printed in the United States of America.

10   9   8   7   6   5   4   3

Contents

Correlation to the New York State Learning Standards and Major Understandings vi

Correlation to the New York City Grade 3 Science Scope and Sequence xiv

Letter to Students xvi

Letter to Parents and Families xvii

What's Ahead in Measuring Up® xviii
  This section provides information about the *New York State Elementary-Level Science Test*, test-taking tips, and strategies for answering multiple-choice questions and open-ended items. You will also learn how to build endurance and confidence to answer those really tough questions that require higher-order skills.

What's Inside: A Lesson Guide xxii

Safety First xxiv
  Find out what standard safety practices should be followed whenever you perform science investigations or activities.

CHAPTER 1 Scientific Inquiry

Standard	NYS Core Curriculum	Lesson		
1, 2	S1.1a, b, S1.2a, S1.3a, IS2.3	1	Science and Inquiry	1
1	S2.1a, S2.3a	2	Lab Safety	5
1, 2	S2.2a, S2.3b, S3.1a, S3.2a, S3.3a, S3.4a, b, IS2.1, IS2.2	3	Analyze and Display Data	9
1	M1.1a, b, c, M2.1a, b	4	Math Skills in Science	13

i

CHAPTER 1 Scientific Inquiry (continued)

Standard	NYS Core Curriculum
1	M3.1a, S2.1a, S2.2a, S2.3a, b
1, 6, 7	PS2.1b, S2.1a, S2.2a, S2.3a, b, S3.1a, S3.2a, S3.3a, S3.4a, ICT6.1, ICT6.5, IPS7.1, IPS7.2

Lesson

5 Tools for Experimenting 17

Performance Task
Graph Temperature Data 21

★ **Building Stamina**® B24

CHAPTER 2 Matter

Standard	NYS Core Curriculum
4	PS3.1a, b, c, d, e
4	PS3.1c, d, e, f, g
4	PS3.2a, b, c
1, 6, 7	S2.1a, S2.2a, T1.2a, T1.4a, ICT6.3, IPS7.2

Lesson

6 Properties of Matter 29
7 Classifying Matter 33
8 States of Matter 37

Performance Task
Plan an Experiment 41

★ **Building Stamina**® B43

CHAPTER 3 Energy

Standard	NYS Core Curriculum
4	PS4.1a, e, f, g
4	PS4.1b, c, e, PS4.2a, b
4	PS4.1b, c, d, f
4	PS4.1c, d, PS4.2a, b
4	PS4.1a, b, c, d, g
1, 7	PS4.1b, c, M3.1a, T1.2a, b, T1.3a, b, c, T1.4a, b, T1.5b, c, ICT6.2, IPS7.1, IPS7.2

Lesson

9 Types of Energy 47
10 Energy Changing Form 51
11 Heat on the Move 55
12 Energy and Matter 59
13 Sound Energy 63

Performance Task
Build Insulated Containers 67

★ **Building Stamina**® B70

CHAPTER 4 Forces and Machines

Standard	NYS Core Curriculum
4	PS5.1a, b, d
4	PS5.1f
4	PS5.1c, e; PS5.2a, b
1, 6, 7	T1.1a, T1.2a, T1.4a, b, T1.5a, b, c, ICT6.2, IPS7.1, IPS7.2

Lesson

14 Forces and Motion.................................... 74

15 Levers, Pulleys, and Inclined Planes 78

16 Gravity and Magnetism............................... 83

Performance Task
  Make a Pulley 88

Building Stamina® **B90**

CHAPTER 5 Plants and Animals

Standard	NYS Core Curriculum
4	LE1.1a, b, c, d, LE2.1a
4	LE2.1a, b, LE2.2a, b
4	LE4.1a, b, c, d, LE4.2a, b, LE5.1a, b
4	LE4.1e, f, g, LE4.2a, b, LE5.1a, b
4	LE1.1b, LE1.2a, LE3.1b, c, LE4.2a, b, LE5.2a, g, LE6.1d, f
4	LE1.1a, LE3.1a, c, LE4.2a, b, LE5.1b, LE5.2b, c, d, e, f, g
4	LE3.2a, b, LE5.2a, b, c, e, f, LE6.1e
1, 4, 6, 7	LE1.1b, LE3.1b, S3.2a, S3.3a, S3.4a, T1.3a, T1.4b, T1.5b, ICT6.6, IPS7.2

Lesson

17 Needs of Living Things 95

18 Traits of Living Things 98

19 Life Cycle of Plants.................................. 104

20 Life Cycle of Animals 109

21 Adaptations of Plants 114

22 Adaptations of Animals.............................. 119

23 Response and Behavior 124

Performance Task
  Water Moves Through Stems 129

Building Stamina® **B132**

iii

CHAPTER 6 Ecology

Standard	NYS Core Curriculum
4	LE5.1a, b, LE6.1a, b, d, LE6.2a
4	LE6.1b, c; LE6.2b
4	LE5.2d, LE5.3a, b, LE7.1a, b, c
1, 6, 7	M2.1a, b, M3.1a, S2.3a, b, S3.2a, S3.4a, b, T1.1b, c, T1.2c, ICT6.1, ICT6.2, ICT6.4, IPS7.2

Lesson

24 Living Things Need Energy 136

25 Food Chains 140

26 Humans and Their Environments 144

Performance Task
   Investigate Air Quality 148

★ **Building Stamina®** B151

CHAPTER 7 Earth

Standard	NYS Core Curriculum
4	PS2.1a, b, c
4	PS2.1c, LE6.2c
4	PS2.1d
4	PS2.1e
4	PS1.1a, b, c
1, 4, 6, 7	PS5.1a, e, PS5.2a, b, T1.4a, b, T1.5a, c, ICT6.2, ICT6.4, IPS7.1, IPS7.2

Lesson

27 Weather 156

28 Water Cycle 160

29 Erosion 164

30 Extreme Natural Events 168

31 Earth, the Moon, and the Sun 173

Performance Task
   Make Your Own Compass 178

★ **Building Stamina®** B181

⭐ **End-of-Book Building Stamina®** B186
It's time for the full review and practice. These challenging, broad-based, higher-level thinking questions will build up your stamina to succeed on other difficult activities.

End-of-Book Performance Test 198

Properties of Common Minerals 208

Commonly Used Units 209

Glossary 210

Measuring Up e-Path® and print DPTs (Diagnostic Practice Tests) .... Measuring Up® Supplements

Your teacher may choose to give Measuring Up e-Path® online formative and summative assessments or print Diagnostic Practice Tests that assess your standards knowledge and identify areas where you need extra support. These assessments will monitor your comprehension of the New York State Learning Standards and help prepare you for the New York State Elementary-Level Science Test.

Explanation of Standards Naming System: Here is how to interpret the New York State Learning Standards used in this Measuring Up® book:

Standard 1
M = Mathematical Analysis   S = Scientific Inquiry   T = Engineering Design

Standard 2
IS = Information Systems

Standard 4
LE = The Living Environment   PS = Physical Setting

Standard 6
ICT = Interconnectedness: Common Themes

Standard 7
IPS = Interdisciplinary Problem Solving

For a correlation of lessons in this Measuring Up® book to the New York City Scope and Sequence for Grade 3, see page xiv.

Correlation to the New York State Learning Standards and Major Understandings

This worktext is customized to the New York *Elementary Science Core Curriculum* and will help you prepare for the *New York State Elementary-Level Science Test* for Grade 4.

New York State Learning Standards and Major Understandings	Measuring Up® Lessons
STANDARDS 1, 2, 6, AND 7: EXPANDED PROCESS SKILLS	
Standard 1—Analysis, Inquiry, and Design Students will use mathematical analysis, scientific inquiry, and engineering design, as appropriate, to pose questions, seek answers, and develop solutions.	
Mathematical Analysis	
M1.1	Use special mathematical notation and symbolism to communicate in mathematics and to compare and describe quantities, express relationships, and relate mathematics to their immediate environment.	
M1.1a	Use plus, minus, greater than, less than, equal to, multiplication, and division signs.	4
M1.1b	Select appropriate operation to solve mathematical problems.	4
M1.1c	Apply mathematical skills to describe the natural world.	4
M2.1	Use simple logical reasoning to develop conclusions, recognizing that patterns and relationships present in the environment assist them in reaching these conclusions.	
M2.1a	Explain verbally, graphically, or in writing the reasoning used to develop mathematical conclusions.	4, Ch 6 PT
M2.1b	Explain verbally, graphically, or in writing patterns and relationships observed in the physical and living environment.	4, Ch 6 PT
M3.1	Explore and solve problems generated from school, home, and community situations, using concrete objects or manipulative materials when possible.	
M3.1a	Use appropriate scientific tools, such as metric rulers, spring scale, pan balance, graph paper, thermometers [Fahrenheit and Celsius], graduated cylinder to solve problems about the natural world.	5, Ch 3, 6 PT
Scientific Inquiry	
S1.1	Ask "why" questions in attempts to seek greater understanding concerning objects and events they have observed and heard about.	
S1.1a	Observe and discuss objects and events and record observations.	1
S1.1b	Articulate appropriate questions based on observations.	1
S1.2	Question the explanations they hear from others and read about, seeking clarification and comparing them with their own observations and understandings.	
S1.2a	Identify similarities and differences between explanations received from others or in print or personal observations and understandings.	1
S1.3	Develop relationships among observations to construct descriptions of objects and events and to form their own tentative explanations of what they have observed.	
S1.3a	Clearly express a tentative explanation or description which can be tested.	1
S2.1	Develop written plans for exploring phenomena or for evaluating explanations guided by questions or proposed explanations they have helped formulate.	
S2.1a	Indicate materials to be used and steps to follow to conduct the investigation and describe how data will be recorded (journal, dates, and times, etc.).	2, 5, Ch 1, 2 PT
S2.2	Share their research plans with others and revise them based on their suggestions.	
S2.2a	Explain the steps of a plan to others, actively listening to their suggestions for possible modification of the plan, seeking clarification and understanding of the suggestions and modifying the plan where appropriate.	3, 5, Ch 1, 2 PT

Ch = Chapter   PT = Performance Task

New York State Learning Standards and Major Understandings		Measuring Up Lessons
S2.3	Carry out their plans for exploring phenomena through direct observation and through the use of simple instruments that permit measurement of quantities, such as length, mass, volume, temperature, and time.	
S2.3a	Use appropriate "inquiry and process skills" to collect data.	2, 5, Ch 1,6 PT
S2.3b	Record observations accurately and concisely.	3, 5, Ch 1,6 PT
S3.1	Organize observations and measurements of objects and events through classification and the preparation of simple charts and tables.	
S3.1a	Accurately transfer data from a science journal or notes to appropriate graphic organizer.	3, Ch 1 PT
S3.2	Interpret organized observations and measurements, recognizing simple patterns, sequences, and relationships.	
S3.2a	State, orally and in writing, any inferences or generalizations indicated by the data collected.	3, Ch 1, 5, 6 PT
S3.3	Share their findings with others and actively seek their interpretations and ideas.	
S3.3a	Explain their findings to others, and actively listen to suggestions for possible interpretations and ideas.	3, Ch 1, 5 PT
S3.4	Adjust their explanations and understandings of objects and events based on their findings and new ideas.	
S3.4a	State, orally and in writing, any inferences or generalizations indicated by the data, with appropriate modifications of their original prediction/explanation.	3, Ch 1, 5, 6 PT
S3.4b	State, orally and in writing, any new questions that arise from their investigation.	3, Ch 6 PT
Engineering Design		
T1.1	Describe objects, imaginary or real, that might be modeled or made differently and suggest ways in which the objects can be changed, fixed, or improved.	
T1.1a	Identify a simple/common object which might be improved and state the purpose of the improvement.	Ch 4 PT
T1.1b	Identify features of an object that help or hinder the performance of the object.	Ch 6 PT
T1.1c	Suggest ways the object can be made differently, fixed, or improved within given constraints.	Ch 6 PT
T1.2	Investigate prior solutions and ideas from books, magazines, family, friends, neighbors, and community members.	
T1.2a	Identify appropriate questions to ask about the design of an object.	Ch 2, 3, 4 PT
T1.2b	Identify the appropriate resources to use to find out about the design of an object.	Ch 3 PT
T1.2c	Describe prior designs of the object.	Ch 6 PT
T1.3	Generate ideas for possible solutions, individually and through group activity; apply age-appropriate mathematics and science skills; evaluate the ideas and determine the best solution; and explain reasons for the choices.	
T1.3a	List possible solutions, applying age-appropriate math and science skills.	Ch 3 PT
T1.3b	Develop and apply criteria to evaluate possible solutions.	Ch 3 PT
T1.3c	Select a solution consistent with given constraints and explain why it was chosen.	Ch 3 PT
T1.4	Plan and build, under supervision, a model of the solution, using familiar materials, processes, and hand tools.	
T1.4a	Create a grade-appropriate graphic or plan listing all the materials needed, showing sizes of parts, indicating how things will fit together, and detailing steps for assembly.	Ch 2, 3, 4, 7 PT
T1.4b	Build a model of the object, modifying the plan as necessary.	Ch 3, 4, 5, 7 PT
T1.5	Discuss how best to test the solution; perform the test under teacher supervision; record and portray results through numerical and graphic means; discuss orally why things worked or didn't work; and summarize results in writing, suggesting ways to make the solution better.	
T1.5a	Determine a way to test the finished solution or model.	Ch 4, 7 PT
T1.5b	Perform the test and record the results, numerically and/or graphically.	Ch 3, 4, 5 PT

Ch = Chapter   PT = Performance Task

New York State Learning Standards and Major Understandings	Measuring Up® Lessons
T1.5c Analyze results and suggest how to improve the solution or model, using oral, graphic, or written formats.	Ch 3, 4, 7 PT
Standard 2—Information Systems Students will access, generate, process, and transfer information using appropriate technologies.	
IS2.1 Information technology is used to retrieve, process, and communicate information and as a tool to enhance learning. • use computer technology, traditional paper-based resources, and interpersonal discussions to learn, do, and share science in the classroom • select appropriate hardware and software that aids in word processing, creating databases, telecommunications, graphing, data display, and other tasks • use information technology to link the classroom to world events	3
IS2.2 Knowledge of the impacts and limitations of information systems is essential to its effectiveness and ethical use. • use a variety of media to access scientific information • consult several sources of information and points of view before drawing conclusions • identify and report sources in oral and written communications	3
IS2.3 Information technology can have positive and negative impacts on society, depending upon how it is used. • distinguish fact from fiction (presenting opinion as fact is contrary to the scientific process) • demonstrate an ability to critically evaluate information and misinformation • recognize the impact of information technology on the daily life of students	1
Standard 6—Interconnectedness: Common Themes Students will understand the relationships and common themes that connect mathematics, science, and technology and apply the themes to these and other areas of learning.	
Systems Thinking	
ICT6.1 Through systems thinking, people can recognize the commonalities that exist among all systems and how parts of a system interrelate and combine to perform specific functions. • observe and describe interactions among components of simple systems • identify common things that can be considered to be systems (e.g., a plant, a transportation system, human beings)	Ch 1, 6 PT
Models	
ICT6.2 Models are simplified representations of objects, structures, or systems, used in analysis, explanation, or design. • analyze, construct, and operate models in order to discover attributes of the real thing • discover that a model of something is different from the real thing but can be used to study the real thing • use different types of models, such as graphs, sketches, diagrams, and maps, to represent various aspects of the real world	Ch 4, 6, 7 PT
Magnitude and Scale	
ICT6.3 The grouping of magnitudes of size, time, frequency, and pressures or other units of measurement into a series of relative order provides a useful way to deal with the immense range and the changes in the scale that affect behavior and design of systems. • observe that things in nature and things that people make have very different sizes, weights, and ages • recognize that almost anything has limits on how big or small it can be	Ch 2 PT
Equilibrium and Stability	
ICT6.4 Equilibrium is a state of stability due to either a lack of changes (static equilibrium) or a balance between opposing forces (dynamic equilibrium). • observe that things change in some ways and stay the same in some ways • recognize that things can change in different ways such as size, weight, color, and movement. Some small changes can be detected by taking measurements	Ch 6, 7 PT
Patterns of Change	

Ch = Chapter   PT = Performance Task

New York State Learning Standards and Major Understandings	Measuring Up® Lessons
ICT6.5 Identifying patterns of change is necessary for making predictions about future behavior and conditions. • use simple instruments to measure such quantities as distance, size, and weight and look for patterns in the data • analyze data by making tables and graphs and looking for patterns of change	Ch 1 PT
Optimization	
ICT6.6 In order to arrive at the best solution that meets criteria within constraints, it is often necessary to make trade-offs. • choose the best alternative of a set of solutions under given constraints • explain the criteria used in selecting a solution orally and in writing	Ch 5 PT
Standard 7—Interdisciplinary Problem Solving Students will understand the relationships and common themes that connect mathematics, science, and technology and apply the themes to these and other areas of learning.	
Connections	
IPS7.1 The knowledge and skills of mathematics, science, and technology are used together to make informed decisions and solve problems, especially those relating to issues of science/ technology/ society, consumer decision making, design, and inquiry into phenomena. • analyze science/ technology/ society problems and issues that affect their home, school, or community, and carry out a remedial course of action. • make informed consumer decisions by applying knowledge about the attributes of particular products and making cost/benefit trade-offs to arrive at an optimal choice. • design solutions to problems involving a familiar and real context, investigate related science concepts to determine the solution, and use mathematics to model, quantify, measure, and compute. • observe phenomena and evaluate them scientifically and mathematically by conducting a fair test of the effect of variables and using mathematical knowledge and technological tools to collect, analyze, and present data and conclusions.	CH 1, 3, 4, 7 PT
Strategies	
IPS7.2 Solving interdisciplinary problems involves a variety of skills and strategies, including effective work habits; gathering and processing information; generating and analyzing ideas; realizing ideas; making connections among the common themes of mathematics, science, and technology; and presenting results. • work effectively • gather and process information • generate and analyze ideas • observe common themes • realize ideas • present results	CH 1–7 PT
SCIENCE SKILLS	
Standard 4—The Physical Setting Students will understand and apply scientific concepts, principles, and theories pertaining to the physical setting and living environment and recognize the historical development of ideas in science.	
PS1.1a Natural cycles and patterns include: • Earth spinning around once every 24 hours (rotation), resulting in day and night • Earth moving in a path around the Sun (revolution), resulting in one Earth ear • the length of daylight and darkness varying with the seasons • weather changing from day to day and through the seasons • the appearance of the Moon changing as it moves in a path around Earth to complete a single cycle	31
PS1.1b Humans organize time into units based on natural motions of Earth: • second, minute, hour • week, month	31
PS1.1c The Sun and other stars appear to move in a recognizable pattern both daily and seasonally.	31
PS2.1a Weather is the condition of the outside air at a particular moment.	27

Ch = Chapter   PT = Performance Task

New York State Learning Standards and Major Understandings	Measuring Up® Lessons
PS2.1b Weather can be described and measured by: • temperature • wind speed and direction • form and amount of precipitation • general sky conditions (cloudy, sunny, partly cloudy)	Ch 1 PT, 27
PS2.1c Water is recycled by natural processes on Earth. • evaporation: changing of water (liquid) into water vapor (gas) • condensation: changing of water vapor (gas) into water (liquid) • precipitation: rain, sleet, snow, hail • runoff: water flowing on Earth's surface • groundwater: water that moves downward into the ground	27, 28
PS2.1d Erosion and deposition result from the interaction among air, water, and land. • interaction between air and water breaks down earth materials • pieces of earth material may be moved by air, water, wind, and gravity • pieces of earth material will settle or deposit on land or in the water in different places • soil is composed of broken-down pieces of living and nonliving earth material	29
PS2.1e Extreme natural events (floods, fires, earthquakes, volcanic eruptions, hurricanes, tornadoes, and other severe storms) may have positive or negative impacts on living things.	30
PS3.1a Matter takes up space and has mass. Two objects cannot occupy the same place at the same time.	6
PS3.1b Matter has properties (color, hardness, odor, sound, taste, etc.) that can be observed through the senses.	6
PS3.1c Objects have properties that can be observed, described, and/or measured: length, width, volume, size, shape, mass or weight, temperature, texture, flexibility, reflectiveness of light.	6, 7
PS3.1d Measurements can be made with standard metric units and nonstandard units. (*Note: Exceptions to the metric system usage are found in meteorology.*)	6, 7
PS3.1e The material(s) an object is made up of determine some specific properties of the object (sink/float, conductivity, magnetism). Properties can be observed or measured with tools such as hand lenses, metric rulers, thermometers, balances, magnets, circuit testers, and graduated cylinders.	6, 7
PS3.1f Objects and/or materials can be sorted or classified according to their properties.	7
PS3.1g Some properties of an object are dependent on the conditions of the present surroundings in which the object exists. For example: • temperature- hot or cold • lighting- shadows, color • moisture- wet or dry	7
PS3.2a Matter exists in three states: solid, liquid, gas. • solids have a definite shape and volume • liquids do not have a definite shape but have a definite volume • gases do not hold their shape or volume	8
PS3.2b Temperature can affect the state of matter of a substance.	8
PS3.2c Changes in the properties or materials of objects can be observed and described.	8
PS4.1a Energy exists in various forms: heat, electric, sound, chemical, mechanical, light.	9, 13
PS4.1b Energy can be transferred from one place to another.	10, 11, 13
PS4.1c Some materials transfer energy better than others (heat and electricity).	10, 11, 12, 13
PS4.1d Energy and matter interact: water is evaporated by the Sun's heat; a bulb is lighted means of electrical current; a musical instrument is played to produce sound; dark colors may absorb light, light colors may reflect light.	11, 12, 13
PS4.1e Electricity travels in a closed circuit.	9, 10

Ch = Chapter   PT = Performance Task

New York State Learning Standards and Major Understandings	Measuring Up® Lessons
PS4.1f Heat can be released in many ways, for example, by burning, rubbing (friction), or combining one substance with another.	9, 11
PS4.1g Interactions with forms of energy can be either helpful or harmful.	9, 13
PS4.2a Everyday events involve one form of energy being changed to another. • animals convert food to heat and motion • the Sun's energy warms the air and water	10, 12
PS4.2b Humans utilize interactions between matter and energy. • chemical to electrical, light, and heat: battery and bulb • electrical to sound (e.g., doorbell buzzer) • mechanical to sound (e.g., musical instruments, clapping) • light to electrical (e.g., solar-powered calculator)	10, 12
PS5.1a The position of an object can be described by locating it relative to another object or the background (e.g., on top of, next to, over under, etc.).	14, Ch 7 PT
PS5.1b The position or direction of motion of an object can be changed by pushing or pulling.	14
PS5.1c The force of gravity pulls objects toward the center of the Earth.	16
PS5.1d The amount of change in the motion of an object is affected by friction.	14
PS5.1e Magnetism is a force that may attract or repel certain materials.	16, Ch 7 PT
PS5.1f Mechanical energy may cause change in motion through the application of force and through the use of simple machines such as pulleys, levers, and inclined planes.	15
PS5.2a The forces of gravity and magnetism can affect objects through gases, liquids, and solids.	16, Ch 7 PT
PS5.2b The force of magnetism on objects decreases as distance increases.	16, Ch 7 PT
Standard 4—The Living Environment Students will understand and apply scientific concepts, principles, and theories pertaining to the physical setting and living environment and recognize the historical development of ideas in science.	
LE1.1a Animals need air, water, and food in order to live and thrive.	17, 22
LE1.1b Plants require air, water, nutrients, and light in order to live and thrive.	17, 21, Ch 5 PT
LE1.1c Nonliving things do not live and thrive.	17
LE1.1d Nonliving things can be human-created or naturally occurring.	17
LE1.2a Living things grow, take in nutrients, breathe, reproduce, eliminate waste, and die.	21
LE2.1a Some traits of living things have been inherited (e.g., color of flowers and number of limbs of animals).	17, 18
LE2.1b Some characteristics result from an individual's interactions with the environment and cannot be inherited by the next generation (e.g., having scars; riding a bicycle).	18
LE2.2a Plants and animals closely resemble their parents and other individuals in their species.	18
LE2.2b Plants and animals can transfer specific traits to their offspring when they reproduce.	18
LE3.1a Each animal has different structures that serve different functions in growth, survival, and reproduction. • wings, legs, or fins enable some animals to seek shelter and escape predators • the mouth, including teeth, jaws, and tongue, enables some animals to eat and drink • eyes, nose, ears, tongue, and skin of some animals enable the animals to sense their surroundings • claws, shells, spines, feathers, fur, scales, and color of body covering enable some animals to protect themselves from predators and other environmental conditions, or enable them to obtain food • some animals have parts that are used to produce sounds and smells to help the animal meet its needs • the characteristics of some animals change as seasonal conditions change (e.g., fur grows and is shed to help regulate body heat; body fat is a form of stored energy and it changes as the seasons change)	22

Ch = Chapter   PT = Performance Task

New York State Learning Standards and Major Understandings		Measuring Up® Lessons
LE3.1b	Each plant has different structures that serve different functions in growth, survival, and reproduction. • roots help support the plant and take in water and nutrients • leaves help plants utilize sunlight to make food for the plant • stems, stalks, trunks, and other similar structures provide support for the plant • some plants have flowers • flowers are reproductive structures of plants that produce fruit which contains seeds • seeds contain stored food that aids in germination and the growth of young plants	21, CH 5 PT
LE3.1c	In order to survive in their environment, plants and animals must be adapted to that environment. • seeds disperse by a plant's own mechanism and/ or in a variety of ways that can include wind, water, and animals • leaf, flower, stem, and root adaptations may include variations in size, shape, thickness, color, smell, and texture • animal adaptations include coloration for warning or attraction, camouflage, defense mechanisms, movement, hibernation, and migration	21, 22
LE3.2a	Individuals within a species may compete with each other for food, mates, space, water, and shelter in their environment.	23
LE3.2b	All individuals have variations, and because of these variations, individuals of a species may have an advantage in surviving and reproducing.	23
LE4.1a	Plants and animals have life cycles. These may include beginning of a life, development into an adult, reproduction as an adult, and eventually death.	19
LE4.1b	Each kind of plant goes through its own stages of growth and development that may include seed, young plant, and mature plant.	19
LE4.1c	The length of time from beginning of development to death of the plant is called its life span.	19
LE4.1d	Life cycles of some plants include changes from seed to mature plant.	19
LE4.1e	Each generation of animals goes through changes in form from young to adult. This completed sequence of changes in form is called a life cycle. Some insects change from egg to larva to pupa to adult.	20
LE4.1f	Each kind of animal goes through its own stages of growth and development during its life span.	20
LE4.1g	The length of time from an animal's birth to its death is called its life span. Life spans of different animals vary.	20
LE4.2a	Growth is the process by which plants and animals increase in size.	19, 20, 21, 22
LE4.2b	Food supplies the energy and materials necessary for growth and repair.	19, 20, 21, 22
LE5.1a	All living things grow, take in nutrients, breathe, reproduce, and eliminate waste.	19, 20, 24
LE5.1b	An organism's external physical features can enable it to carry out life functions in its particular environment.	19, 20, 24
LE5.2a	Plants respond to changes in their environment. For example, the leaves of some green plants change position as the direction of light changes; the parts of some plants undergo seasonal changes that enable the plant to grow; seeds germinate, and leaves form and grow.	21, 23
LE5.2b	Animals respond to change in their environment (e.g., perspiration, heart rate, breathing rate, eye blinking, shivering, and salivating).	22, 23
LE5.2c	Senses can provide essential information (regarding danger, food, mates, etc.) to animals about their environment.	22, 23
LE5.2d	Some animals, including humans, move from place to place to meet their needs.	22, 26
LE5.2e	Particular animal characteristics are influenced by changing environmental conditions including: fat storage in winter, coat thickness in winter, camouflage, shedding of fur.	22, 23
LE5.2f	Some animal behaviors are influenced by environmental conditions. These behaviors may include: nest building, hibernating, hunting, migrating, and communicating.	22, 23
LE5.2g	The health, growth, and development of organisms are affected by environmental conditions such as the availability of food, air, water space, shelter, heat, and sunlight.	21, 22

Ch = Chapter   PT = Performance Task

New York State Learning Standards and Major Understandings		Measuring Up® Lessons
LE5.3a	Humans need a variety of healthy foods, exercise, and rest in order to grow and maintain good health.	26
LE5.3b	Good health habits include hand washing and personal cleanliness; avoiding harmful substances (including alcohol, tobacco, illicit drugs); eating a balanced diet; engaging in regular exercise.	26
LE6.1a	Green plants are producers because they provide the basic food supply for themselves and animals.	24
LE6.1b	All animals depend on plants. Some animals (predators) eat other animals (prey).	24, 25
LE6.1c	Animals that eat plants for food may in turn become food for other animals. This sequence is called a food chain.	25
LE6.1d	Decomposers are living things that play a vital role in recycling nutrients.	21, 24
LE6.1e	An organism's pattern of behavior is related to the nature of that organism's environment, including the kinds and numbers of other organisms present, the availability of food and other resources, and the physical characteristics of the environment.	23
LE6.1f	When the environment changes, some plants and animals survive and reproduce, and others die or move to new locations.	21
LE6.2a	Plants manufacture food by utilizing air, water, and energy from the Sun.	24
LE6.2b	the Sun's energy is transferred on Earth from plants to animals through the food chain.	25
LE6.2c	Heat energy from the Sun powers the water cycle.	28
LE7.1a	Humans depend on their natural and constructed environments.	26
LE7.1b	Over time humans have changed their environment by cultivating crops and raising animals, creating shelter, using energy, manufacturing goods, developing means of transportation, changing populations, and carrying out other activities.	26
LE7.1c	Humans, as individuals or communities, change environments in ways that can be either helpful or harmful for themselves and other organisms.	26

Ch = Chapter   PT = Performance Task

Correlation to the New York City Grade 3 Science Scope and Sequence

This worktext is customized to the New York City *Science Scope and Sequence* and will help you prepare for the *New York City Science Assessment* for Grade 3.

NYC Scope and Sequence	NYS Learning Standard(s)	Measuring Up® Lessons
UNIT 1: Matter		
What are some of the properties of matter?		
Measure, compare and record physical properties of objects using: • Standard (metric) and nonstandard units • Appropriate tools (e.g., rulers, thermometers, pan balances, spring scales, graduated cylinders, beakers)	PS 3.1b, PS 3.1c, PS 3.1d, PS 3.1e	6, 7
Describe and compare the physical properties of matter (size, shape, mass/weight, volume, flexibility, luster, color, texture, hardness, odor, etc.)	PS 3.1b, PS 3.1c	6, 7
UNIT 2: Energy		
What are some ways that energy can be changed from one form to another?		
Observe, identify, and describe a variety of forms of energy: • Sound • Heat • Chemical • Mechanical • Electricity	PS 4.1a	9, 13
Identify the evidence for energy transformations and how humans use these energy transformations: • Heat to light, chemical to electrical, electrical to sound, etc.	PS 4.2a, PS 4.2b	10, 12
Observe and describe how heat is conducted and can be transferred from one place to another.	PS 4.1b, PS 4.1c, PS 4.1d	10, 11, 12, 13, Ch 3 PT
Observe and describe different ways in which heat can be released: • Burning, rubbing (friction), or combining one substance with another.	PS 4.1f	9
Interactions of matter and energy (e.g., electricity lighting a bulb, dark colors absorbing light, etc).	PS 4.1d	11, 12, 13
Sound energy: • Pitch (frequency) • Vibrations • Volume • How sound travels through solids, liquids, gases • Noise pollution	PS 4.1a, PS 4.1b, PS 4.1c, PS 4.1d, PS 4.1g	9, 10, 11, 12, 13, Ch 3 PT
UNIT 3: Simple Machines		
How do simple machines help us move objects?		
Demonstrate how mechanical energy may cause change in motion through the application of force or the use of simple machines, such as: • Levers, pulleys, inclined planes • Wheel and axle	PS 5.1f	15
Observe and describe how the amount of change in the motion of an object is affected by friction.	PS 5.1d	14
Observe and describe how the position or direction of motion of an object can be changed by pushing or pulling.	PS 5.1b	14
Observe how the force of gravity pulls objects toward the center of the Earth.	PS 5.1c	16

NYC Scope and Sequence	NYS Learning Standard(s)	Measuring Up® Lessons
UNIT 4: Plant and Animal Adaptations		
How are plants and animals well-suited to live in their environments?		
Describe how all living things grow, take in nutrients, breathe, reproduce and eliminate wastes.	LE 5.1a, LE 5.1b	19, 20, 24
Describe how plants must be adapted to their environment in order to survive: • Structures and their functions (e.g., roots, leaves, flowers, etc.) • Adaptations of these structures may include variations in size, shape, thickness, color, smell, and texture • Plants change as the seasons change • See dispersal	LE 3.1b, LE 3.1c, LE 5.2a, LE 6.1f	21, 22, 23, 26, Ch 5 PT

Describe how animals must be adapted to their environment in order to survive:
• Structures and functions (e.g., wings, legs, fins, scales, feathers, fur, etc.)
• Understand that animals respond to change in the environment (e.g., heart rate, eye blinking, shivering)
• Animals change as seasons change
  ◦ Hibernation
  ◦ Migration (i.e., moving from place to place to meet needs) including human
	LE 3.1a, LE 3.1c, LE 5.2b, LE 5.2d, LE 5.2e, LE 5.2f, LE 6.1f	21, 22, 23, 26

Recognize that traits of living things are both: • Inherited (color of flowers, eye color) • Learned/acquired (riding a bicycle, having scars)	LE 2.1a, LE 2.1b	17, 18

Dear Student:

How do you get better at anything you do? You practice! Just as with sports or other activities, the key to success in school is practice, practice, practice.

This book will help you review and practice science skills. These are the skills you need to know to measure up to the New York State Learning Standards for Science for your grade. Practicing these skills now will help you do better in your work all year.

This Measuring Up® book has 7 chapters. Each chapter gives you practice in using your thinking skills.

Each lesson consists of four main sections:

- **Focus on the NYS Learning Standards** introduces the NYS Learning Standard skills covered in the lesson.
- **Guided Instruction** guides you through a review of science concepts and skills you will need for successful learning.
- **Apply the NYS Learning Standards** helps you practice NYS Learning Standard skills you have been learning with short-answer questions.
- **NYS Test Practice** gives you practice in answering test-type questions.

In addition to the lessons, other sections in the book are called **Building Stamina**®. These sections contain multiple-choice and short-response questions that help build your intellectual brainpower. Many of these questions are more difficult and will help you prepare for taking tests.

There is also a section in each chapter called **Performance Task**. In this section, you will be performing short experiments based on skills you have learned throughout that particular chapter.

In the fourth grade, you will take the *New York State Elementary-Level Science Test*. It will be an important step forward. The NYS Test will show how well you measure up to the New York State Learning Standards. It is just one of the many important tests you will take. Success on the NYS Test will prepare you for the next level of science challenges.

Have a great and successful year!

New York State Learning Standards
and Success Strategies for the State Test

To Parents and Families,

All students need science knowledge and skills to succeed. New York educators have created standards called the New York State Learning Standards for Science. The NYS Learning Standards describe what all New York students should know at each grade level. Students need to meet these standards to graduate, as measured by the *New York State Elementary-Level Science Test*.

The NYS Test is directly related to the Learning Standards. The NYS Test emphasizes higher-level thinking skills. Students must learn to consider, analyze, interpret, and evaluate instead of just recalling simple facts.

Measuring Up® will help your child to review the Learning Standards and prepare for all science exams. It contains:

- **Lessons** that focus on practicing the NYS Learning Standards;
- **Guided Instruction**, in which students are shown the steps and skills necessary to solve a variety of science problems;
- **Apply the NYS Learning Standards**, which provides independent practice with concepts and skills reviewed in the lesson;
- **NYS Test Practice**, which shows how individual standards can be understood through multiple-choice and short-response questions.
- **Building Stamina**®, which gives practice with more difficult multiple-choice and short-response questions that require higher-level thinking.
- **Performance Tasks**, which gives practice in performing classroom experiments using short-response questions, data tables, and creating drawings.

For success in school and the real world, your child needs to be successful in science. Get involved! Your involvement is crucial to your child's success. Here are some suggestions:

- Involve your child in activities that require science concepts and skills, such as mixing recipes, exploring the ecology of your neighborhood and community, observing and studying the night sky, and recycling.
- Help to find appropriate Internet sites for science.
- Note how science is used when you are out with your family. Discuss how science is used in preparing meals, in careers such as medicine and architecture, in space exploration, and in other real-life applications.
- Encourage your child to talk about what he or she has learned in science class.

Work this year to ensure your child's success. Science skills are essential for success and throughout your child's life.

What's Ahead in Measuring Up

This book was created for New York students like you. Each lesson, question, and Performance Task is aimed at helping you master the Learning Standards and do well on the NYS Test. It will also help you do well on other science tests you take during the school year.

About the NYS Test

New York educators have set up standards for science. They are called the New York State Learning Standards for Science. They spell out what all students at each grade level should know. New York educators have also created a statewide test for science. It is called the *New York State Elementary-Level Science Test*. This test measures how well students have mastered the standards.

Format of the NYS Test

The *New York State Elementary-Level Science Test*, which you will take in the fourth grade, has two types of test items:

- multiple-choice questions
- open-ended questions

Many questions include a picture, diagram, chart, or other type of graphic, which is used to answer the question. Measuring Up® gives you practice in reading and using these types of graphics.

Measuring Up® on Multiple-Choice Questions

You are probably familiar with the multiple-choice type of question. It has a question, or stem, followed by four answer choices. Your job is to select the one correct choice. On the NYS Test, you will answer many multiple-choice questions. Here are some strategies:

- Always try to determine the answer without looking at the choices. Once you arrive at an answer, compare your answer with the choices.
- If your answer is not one of the choices, check your work and rethink your ideas carefully.
- Some multiple-choice questions refer to a graphic such as an illustration, a graph, a table, or a picture. You will be asked to read or interpret the graphic. Read the question carefully and use the graphic to answer the specific question.

- Many questions test higher-order thinking skills. You must connect the ideas and information to come up with the right answers.
- Even if you don't know the answer to a multiple-choice question, you can make a good guess based on what you know and get the question right.
- Check and double-check your answers before you turn in the test. Be sure of your answers and be sure you haven't marked a wrong answer choice by mistake.

Measuring Up® on Open-Ended Questions

Open-ended questions are simply questions that require short written answers. When you see one of these questions, you will be asked to write your answer on the lines provided. Or, your teacher may ask you to write your answers in your science notebook. Here are some tips:

- Carefully study the information and the questions. Because you do not have answer choices as a way to check yourself, it is important to take your time and follow all the steps carefully.
- Often you will be required to study a graphic or read a short paragraph. Use the information provided to answer the question.
- Once you have an answer, carefully write it on the lines provided. You may need to sketch part of your answer. Reread your answer to make sure it is clear and it says what you want it to say.

Measuring Up® with Building Stamina®

A special feature of Measuring Up® is **Building Stamina®**, designed to give you practice and build your confidence and endurance for completing higher-level thinking activities. These activities include answering questions that cover multiple Learning Standards. Each chapter ends with a **Building Stamina®** section. At the end of the book is a longer, comprehensive **Building Stamina®**, which is a complete review of all the standards covered in the lessons.

Measuring Up® with Performance Tests

A special feature at the end of each chapter is called the **Performance Task**. These are designed to help you practice for the *New York State Elementary-Level Science Test*.

At the end of this book is the **End-of-Book Performance Test**. With three separate stations, the Performance Test is designed to be just like the real *New York State Elementary-Level Science Test* that you will take in the fourth grade. This will help build your confidence and endurance for completing classroom experiments and activities. Here are some directions on taking this test:

- The Performance Test consists of hands-on tasks set up at three stations.
- When you take the state test, each of the performance tasks will be timed. In the End-of-Book Performance Test, you will have 15 minutes to complete the tasks at each of the three stations.
- The Performance Test is composed of all open-ended items assessing your skills in using hands-on equipment and materials in your responses to the questions posed (primarily from Standard 1 of the NYS Learning Standards).
- Detailed directions for each station will be provided within the station itself. Read these directions carefully.
- Make sure you complete all data tables, graphs, and questions. You will be scored on how complete and accurate your responses are.

Higher-Level Thinking Skills ★

The NYS Test is designed to tap your higher-level thinking skills. When you use higher-level thinking skills, you do more than just recall information. For instance, instead of being asked to name the parts of the circulatory system, you might need to describe how this system functions and affects the body of a living organism. Higher-level thinking questions are starred in the NYS Test Practice pages and **Building Stamina**® sections of this book.

Tips to Measure Up

Keep these general tips in mind to help you prepare for the test:

- Start preparing now. Pace yourself. Spend a few minutes a day practicing answering test questions. Right now, the test may seem far in the future, but you will be sitting with the NYS Test in front of you before you know it.

- Get a good night's sleep the night before the test. Do not expect to cram everything into your head the night before. You can't remember much that way, and you will be too tired to do well.

- Eat a good breakfast. If you are hungry during the test, you will be distracted and unable to think clearly.

- Think positively. Do not focus on the things you do not know. Nobody is perfect. If you are unsure of an answer, mark that question and move on to the next. After you have worked through the test, return to those questions you have marked.

You will learn a lot in Measuring Up®. You will review and practice the NYS Learning Standards. You will practice for the NYS Test. Finally, you will build your stamina to answer tough questions. You will more than measure up. You'll be a smashing success!

What's Inside: A Lesson Guide

The lessons in this worktext first introduce individual learning standards and then explain, apply, and assess the concepts and skills that are needed to meet those standards.

Focus on the NYS Learning Standards
Introduces the NYS Learning Standards and important terms and concepts covered in the lesson. These terms are defined in a glossary at the end of the book.

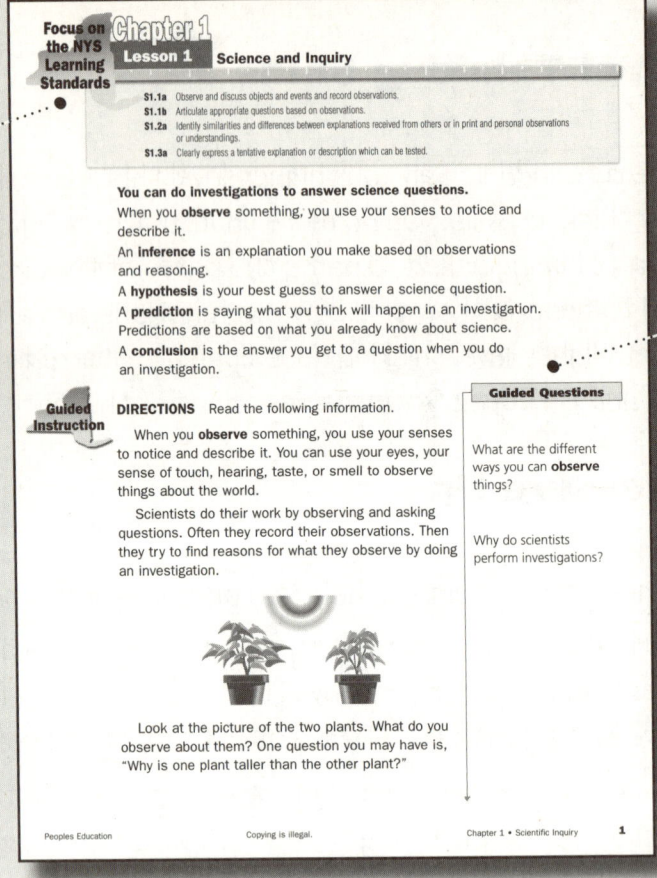

Guided Instruction
Work with your teacher to learn and review important science concepts.

Apply the NYS Learning Standards
Provides practice for the important concepts and skills learned in the "Guided Instruction" section of the lesson.

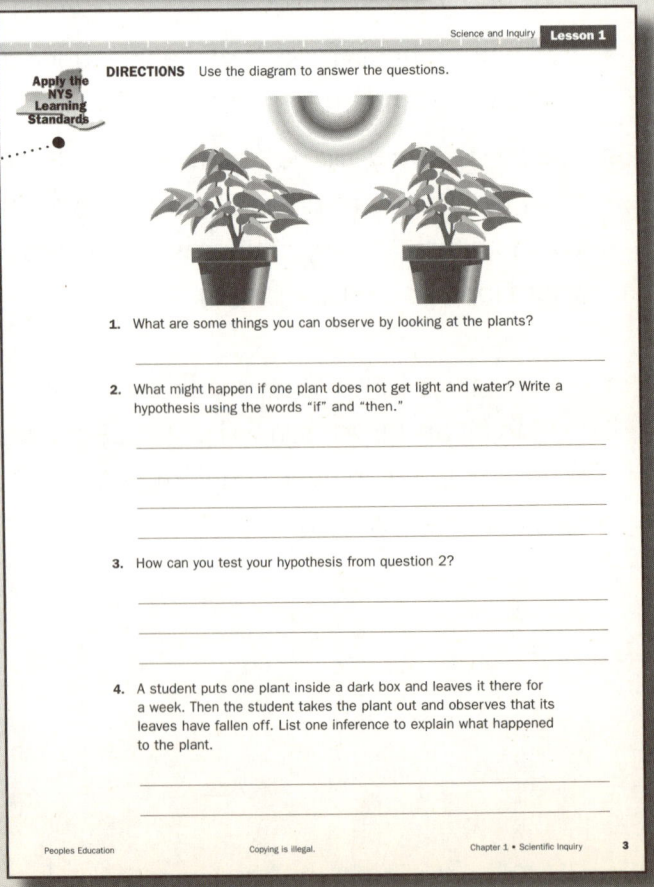

NYS Test Practice
Gives multiple-choice and open-ended questions that will test the understanding of the concepts and skills taught in the lesson.

Lesson 1 — Science and Inquiry

NYS Test Practice

DIRECTIONS Choose the best answer for each question. Then circle the letter of the answer you have chosen.

1. What kind of statement is "The rock feels bumpy and scratchy"?
   A hypothesis
   B prediction
   C observation
   D inference

2. A student wants to know if beans planted in a warm place will grow better than beans planted in a cool place. Which is the best hypothesis for the student to use?
   A If the seeds are kept warm, then they will grow faster.
   B If the beans are planted in good soil, then they will grow faster.
   C If the beans are given enough water, then they will grow faster.
   D If the beans are given fertilizer, then they will grow faster.

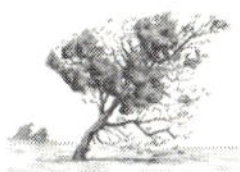

3. Look at the picture above. Which is the best question to ask about the picture?
   A What kind of flower is this?
   B What is causing the tree to lean?
   C What color is the bark?
   D The tree has leaves on it.

4. A student planted the same kind of flower seeds in two pots. The student added plant food to one pot. The seeds in this pot grew faster than the other seeds. What is the best conclusion?
   A Flower seeds need plant food to grow.
   B Flower seeds do not need plant food to grow.
   C Some flower seeds grow better than others.
   D Plant food helps the flower seeds grow faster.

Performance Tasks
Work with your teacher to do science investigations. These will help you understand important science concepts. Remember to follow your teacher's directions and to follow safety rules.

Performance Task
Graph Temperature Data

Focus on the NYS Learning Standards: PS2.1b; S2.1a; S2.2a; S2.3a; S2.3b; S3.1a; S3.2a; S3.3a; S3.4a; ISC6.1; IPS7.1; IPS7.2

Task:
You will make a bar graph to show average monthly temperatures in Yonkers, New York.

Materials:
- pencil
- red crayon, colored pencil, or marker
- yellow crayon, colored pencil, or marker
- blue crayon, colored pencil, or marker

Procedure:
- Look at the table. Then answer questions 1 and 2.
- Complete a bar graph. Use the information in the table to complete your graph. Use the materials above to complete this task.

Month	Average Temperature in Degrees Celsius
January (Jan)	3
February (Feb)	6
March (Mar)	11
April (Apr)	17
May (May)	22
June (Jun)	27
July (Jul)	30
August (Aug)	28
September (Sep)	24
October (Oct)	18
November (Nov)	12
December (Dec)	6

SAFETY FIRST

This book contains various investigations and activities that demonstrate the concepts in Measuring Up® *to the New York State Learning Standards*. Following standard safety practices is an important laboratory procedure when completing any science activity.

Before You Begin This Book

1. Review this page of safety guidelines.

2. Always follow your teacher's directions.

Before You Experiment

3. Make sure your teacher or another adult is present to supervise your work.

4. Read the instructions for each science activity before you begin.

5. Wear the safety equipment that your teacher tells you to. If your hair is long, tie your hair back.

During an Experiment

6. Follow the instructions step-by-step in the order that they are presented.

7. Never run in a lab or play games during an experiment.

8. Do not bring food or drink into the lab or classroom.

9. Check to see that all containers are labeled so you know what substances they hold.

10. Substances used in experiments can be dangerous. Only taste them or smell them if your teacher tells you to.

11. Mix ingredients only as your activity instructs. Playing with these ingredients may create dangerous substances.

12. Remember, knives and scissors are sharp. Move the knife or scissors away from your body when you are cutting.

13. Accidents do occur. Someone may be hurt or something may be broken. Immediately tell your teacher or the adult supervising your work.

After the Experiment Is Done

14. Ask your teacher what to do with unused ingredients and containers.

15. Follow your teacher's instructions to clean up your work area.

16. Make sure you turn off all lights, switches, burners, and faucets.

Chapter 1
Lesson 1   Science and Inquiry

Focus on the NYS Learning Standards

- **S1.1a** Observe and discuss objects and events and record observations.
- **S1.1b** Articulate appropriate questions based on observations.
- **S1.2a** Identify similarities and differences between explanations received from others or in print and personal observations or understandings.
- **S1.3a** Clearly express a tentative explanation or description which can be tested.
- **IS2.3** Information technology can have positive positive and negative impacts on society, depending upon how it is used.

You can do investigations to answer science questions.

When you **observe** something, you use your senses to notice and describe it.

An **inference** is an explanation you make based on observations and reasoning.

A **hypothesis** is your best guess to answer a science question.

A **prediction** is saying what you think will happen in an investigation. Predictions are based on what you already know about science.

A **conclusion** is the answer you get to a question when you do an investigation.

DIRECTIONS   Read the following information.

When you **observe** something, you use your senses to notice and describe it. You can use your eyes, your sense of touch, hearing, taste, or smell to observe things about the world.

Scientists do their work by observing and asking questions. Often they record their observations. Then they try to find reasons for what they observe by doing an investigation.

Guided Questions

What are the different ways you can **observe** things?

Why do scientists perform investigations?

Look at the picture of the two plants. What do you observe about them? One question you may have is, "Why is one plant taller than the other plant?"

Lesson 1 — Science and Inquiry

You know that plants need water, air, light, and a place to grow. You can reason that the shorter plant does not get enough water. This idea is called an inference. An **inference** is an idea you get to explain something. You make inferences using observations and reasoning.

Asking questions and making inferences help you to make a hypothesis. A **hypothesis** is your best guess to answer a science question. You can write a hypothesis using the words "if" and "then." Your hypothesis could be, "If a plant is watered every day, then it will grow taller."

A hypothesis must be tested to see if it is right. To test your hypothesis, you can do an investigation. You can give two identical plants different amounts of water, every day. At the same time, you can measure and write down how tall both plants are.

A **prediction** is what you think will happen in an investigation. We make predictions using what we already know. In your hypothesis, you predict that a plant will be taller if it gets water every day. The measurements from your investigation will show if this prediction is right or wrong.

As you do an investigation, you write down the things you observe. When the investigation is over, you can read your observations and make a conclusion. A **conclusion** is the answer you get to your questions from your investigation.

Guided Questions

What are some other questions you might ask about the plants?

What is a hypothesis?

What is a conclusion?

DIRECTIONS For each question, write your answer in the spaces provided.

1. What does it mean to observe something?

   _____

2. What is an inference?

   _____

3. Are predictions for an investigation always right? Explain.

   _____

   _____

Science and Inquiry | **Lesson 1**

Apply the NYS Learning Standards

DIRECTIONS Use the diagram to answer the questions.

1. What are some things you can observe by looking at the plants?

   _____

2. What might happen if one plant does not get light and water? Write a hypothesis using the words "if" and "then."

   _____
   _____
   _____
   _____

3. How can you test your hypothesis from question 2?

   _____
   _____
   _____

4. A student puts one plant inside a dark box and leaves it there for a week. Then the student takes the plant out and observes that its leaves have fallen off. List one inference to explain what happened to the plant.

   _____
   _____

Peoples Education · Copying is illegal. · Chapter 1 • Scientific Inquiry

Lesson 1 Science and Inquiry

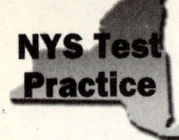

DIRECTIONS Choose the best answer for each question. Then circle the letter of the answer you have chosen.

1. What kind of statement is "The rock feels bumpy and scratchy"?
   A hypothesis
   B prediction
   C observation
   D inference

2. A student wants to know if beans planted in a warm place will grow better than beans planted in a cool place. Which is the best hypothesis for the student to use?
   A If the seeds are kept warm, then they will grow faster.
   B If the beans are planted in good soil, then they will grow faster.
   C If the beans are given enough water, then they will grow faster.
   D If the beans are given fertilizer, then they will grow faster.

3. Look at the picture above. Which is the best question to ask about the picture?
   A What kind of flower is this?
   B What is causing the tree to lean?
   C What color is the bark?
   D The tree has leaves on it.

4. A student planted the same kind of flower seeds in two pots. The student added plant food to one pot. The seeds in this pot grew faster than the other seeds. What is the best conclusion?
   A Flower seeds need plant food to grow.
   B Flower seeds do not need plant food to grow.
   C Some flower seeds grow better than others.
   D Plant food helps the flower seeds grow faster.

Lesson 2 — Lab Safety

S2.1a Indicate materials to be used and steps to follow to conduct the investigation and describe how data will be recorded (journal, dates and times, etc.).

S2.3a Use appropriate "inquiry and process skills" to collect data.

It is important for you to follow safety rules when doing a scientific activity.

A **precaution** is something done to keep an accident from happening.

A **hazard** is something that is dangerous.

Guided Instruction

DIRECTIONS Read the following information.

Sometimes you will do science activities or investigations in the classroom. Other times you will do these science activities outdoors. Science is fun but you want to be safe when you do science activities. You can be safe by following safety guidelines, or rules.

1. Follow your teacher's directions.
2. Take precautions, or safety steps, before you begin the activity. Taking precautions will help protect you from hazards. Use safety goggles to protect your eyes. Use gloves and lab aprons to protect you from spills. Tie back long hair and roll up loose sleeves to avoid accidents.

Guided Questions

What precautions are shown in the picture?

Lesson 2 — Lab Safety

3. Read the directions before you start. Ask your teacher to explain anything you do not understand. Follow the steps of an activity in order.

4. Be careful of the hazards, or dangers, in science class. Tell your teacher when an accident happens. Tell your teacher when something spills or breaks.

5. Be sure to keep your work area neat. Put away anything you do not need. You might knock something over if your work area is messy.

6. Do not taste things or put your hands into your mouth during science activities. Keep food and drink away from your work area.

7. Clean up your work area. Put away equipment. Ask your teacher what to do with extra materials. Be sure to wash your hands with soap and water.

8. Follow rules when you do a science activity outside. Stay together with your group. Only touch plants and animals your teacher asks you to touch.

Guided Questions

What should you do if you are not sure what to do during a science activity?

What is a hazard?

What hazard is shown in the picture on this page?

What should you keep away from your work area?

DIRECTIONS For each question, write your answer in the spaces provided.

1. What should you do before you start a science activity?

   _____
   _____

2. What should you do during a science activity so that your results will be correct?

   _____
   _____

3. What are some precautions you can follow to protect yourself?

   _____
   _____

Lab Safety — Lesson 2

Apply the NYS Learning Standards

DIRECTIONS Read the text below and answer the questions.

Today, you and your classmates are going to do a science activity. You read the directions. Then your teacher asks if anyone has questions about it. Everyone puts on safety goggles, gloves, and lab aprons. You follow each step in order. You mix liquid soap, vinegar, and baking soda. You watch what happens. When you finish, you clean your work area and wash your hands.

1. Why is it important to read the activity before beginning?

   _____
   _____
   _____

2. What might happen if you do not follow the steps of the activity in order?

   _____
   _____

3. Why does everyone wear goggles, gloves, and lab aprons during the science activity?

   _____
   _____

4. What should you do if you spill something?

   _____
   _____

Lesson 2 — Lab Safety

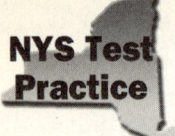

NYS Test Practice

DIRECTIONS Choose the best answer for each question. Then circle the letter of the answer you have chosen.

1. You want to do a science activity on your own. You should first
   - A follow all safety guidelines while doing it
   - B read the activity and ask questions
   - C do the activity alone
   - D ask your teacher if you can do the activity

2. Which of the following is *not* a hazard shown in the picture?

   - A messy work area
   - B goggles being worn
   - C apple on the desk
   - D hair not tied back

3. Which of the following is a precaution?
   - A Read the activity and ask questions.
   - B Put away all equipment.
   - C Wear safety goggles and roll up loose sleeves.
   - D Wash your hands.

4. 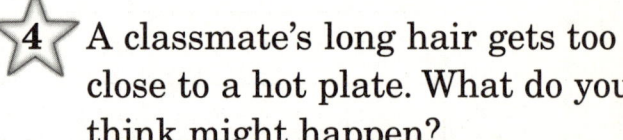 A classmate's long hair gets too close to a hot plate. What do you think might happen?
   - A Her hair might get burned.
   - B Goggles will protect her hair.
   - C The hot plate will turn itself off.
   - D The hot plate will break.

5. The best time to remove your safety goggles and lab apron is
   - A during the activity
   - B after your work area is clean and all equipment is put away
   - C when the first person removes his or hers
   - D when you get tired of wearing them

Lesson 3 — Analyze and Display Data

Focus on the NYS Learning Standards

- **S2.3b** Record observations accurately and concisely.
- **S3.1a** Accurately transfer data from a science journal or notes to appropriate graphic organizer.
- **S3.2a** State, orally and in writing, any inferences or generalizations indicated by the data collected.
- **S3.3a** Explain their findings to others, and actively listen to suggestions for possible interpretations and ideas.
- **S2.2a, S3.4a, b, IS2.1, IS2.2**

You can use charts to write down and arrange information. You can also show information in a graph to help you understand it.

Data are the facts gathered from an investigation.

A **bar graph** uses bars to show different amounts.

A **picture graph** uses symbols or pictures to show amounts.

A **line plot** shows data along a number line.

Guided Instruction

DIRECTIONS Read the following information.

Data are the observations and information gathered from an investigation. It is important to write down your data clearly and carefully or enter the information into a computer program for analysis. You can put your data into a chart or table. If you use a variety of source materials, it will be important to keep track of them. The title of the table tells what kind of data it shows. Each column in a table also has a title. Look at the table below.

TYPE AND NUMBER OF TREES IN A PARK

Type of Tree	Number of Trees
Oak	4
Maple	8
Pine	2

After you record data in a table, you can make a graph. Graphs help you understand data. A bar graph uses bars to show data. You can put the data from the table above into a bar graph. The graph helps you to quickly see that most of the trees in the park are maple trees.

Guided Questions

What kind of information is written in charts?

Why are graphs used to show data?

Lesson 3 — Analyze and Display Data

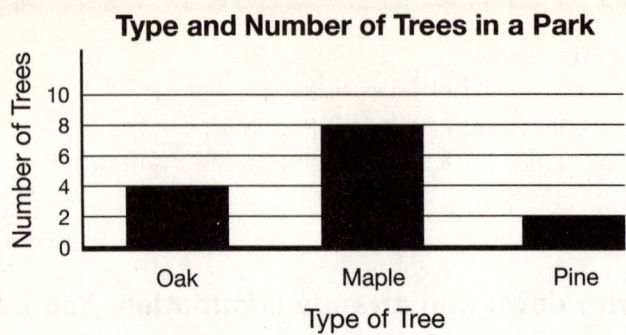

Guided Questions

What does each bar on the bar graph stand for?

A picture graph uses pictures or symbols to show amounts. Each picture may stand for one or more of the same thing. Look at the key on the picture graph below.

Type of Trees

Oak	🌲🌲
Maple	🌲🌲🌲🌲
Pine	🌲

Key: Each 🌲 means 2 trees.

How many oak trees are shown on the picture graph?

A line plot is a graph that shows data along a number line. This line plot shows the ages of the maple trees in the park. Each x on the number line stands for one tree.

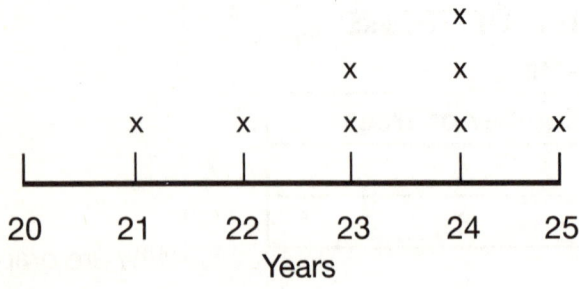

In the line plot, how many maple trees are 23 years old?

DIRECTIONS For each question, write your answer in the spaces provided.

1. What is a line plot?

___

2. What is a bar graph?

___

Analyze and Display Data   Lesson 3

Apply the NYS Learning Standards

DIRECTIONS Use the graph to answer the questions.

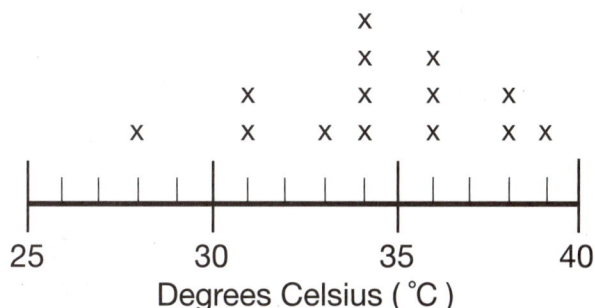

Highest Daily Temperature for Two Weeks

Degrees Celsius (°C)

1. What type of graph is shown?

   _____

2. What do the numbers on the bottom of the graph stand for?

   _____

3. What units of temperature measurement are used?

   _____

4. How many days had a high temperature of 39°C?

   _____

5. What is the lowest temperature shown for the two weeks?

   _____

6. Which temperature was reached the most number of times?

   _____

   _____

7. Why is it important to record data carefully and correctly?

   _____

   _____

   _____

Lesson 3 Analyze and Display Data

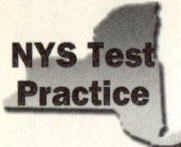

NYS Test Practice

DIRECTIONS Choose the best answer for each question. Then circle the letter of the answer you have chosen.

1. What is the information gathered from an investigation called?

   A axis

   B unit

   C data

   D trend

Books Read by Four Students

Cristina	📕📕📕📕
Robert	📕📕
Ginny	📕
Carlos	📕📕📕

Key: 📕 means 5 books.

 2. Use the graph above to find the number of books Carlos read.

   A 3

   B 6

   C 10

   D 15

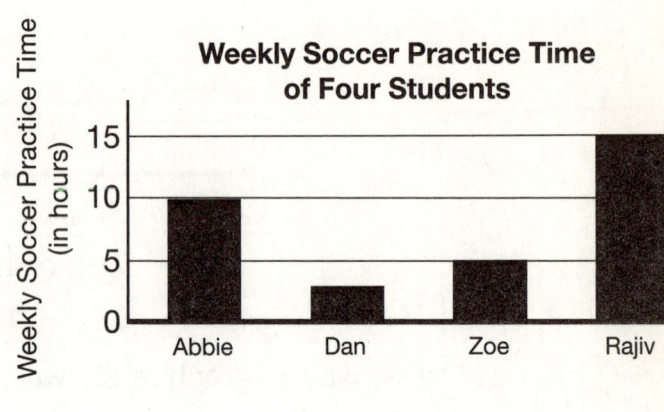

3. What is the best title for the missing label in the bar graph above?

   A Day

   B Week

   C Hours

   D Students

 4. What does the bar graph above show?

   A Rajiv practices the most hours each week.

   B Abbie practices the most hours each week.

   C Dan practices more each week than Zoe.

   D Rajiv practices less each week than Abbie.

Focus on the NYS Learning Standards

Lesson 4 — Math Skills in Science

M1.1a Use plus, minus, greater than, less than, equal to, multiplication, and division signs.
M1.1b Select the appropriate operation to solve mathematical problems.
M1.1c Apply mathematical skills to describe the natural world.
M2.1a Explain verbally, graphically, or in writing the reasoning used to develop mathematical conclusions.
M2.1b Explain verbally, graphically, or in writing patterns and relationships observed in the physical and living environment.

You can use math skills to describe the natural world.

If you have good **math skills**, you can choose the correct operation to solve a problem. In math, **operations** include addition, subtraction, multiplication, and division.

Guided Instruction

DIRECTIONS Read the following information.

You know a science investigation begins with a question. One question might be: Does the temperature always drop after it rains?

To answer a science question, you must make observations. You must record data. After that, you often must use **math skills** to describe your observations and data.

To answer the question above, you might record the temperatures before and after several rainfalls. You might put the data in a table like this one.

Date of Rainfall	Temperature (°F) Before the Rain	Temperature (°F) After the Rain
May 20	68°F	59°F
May 28	70°F	64°F
June 3	72°F	65°F
June 4	64°F	62°F

Once you have the data, you must explain what it means. Here you will compare the temperatures before and after the rainfall. To compare, you must choose an **operation** to use. Will you add, subtract, multiply, or divide? In this case, you will subtract.

May 20: 68 − 59 = 9°F
May 28: 70 − 64 = 6°F
June 3: 72 − 65 = 7°F
June 4: 64 − 62 = 2°F

Guided Questions

How might **math skills** help a scientist?

What **operation** do you use to compare two numbers?

Peoples Education — Copying is illegal. — Chapter 1 • Scientific Inquiry

Lesson 4 Math Skills in Science

Subtraction helps you see the difference between the temperatures before and after the rainfall. It helps you see that after a rainfall the temperature did drop for these four days.

Checking the temperature change for only four days is not enough to draw a conclusion. It could happen that the temperature would not go down the next time it rained. You would need many more observations to draw a good conclusion. Still, the math skills you've used can help you describe what you saw.

Scientists use all four operations to work on science problems. Here are some examples.

OPERATION	PROBLEM	NUMBER SENTENCE
Addition	2 chickadees, 4 goldfinches, and 1 nuthatch visited the feeder in one hour. How many birds in all visited the feeder?	$2 + 4 + 1 = 7$ birds
Subtraction	Rocket A flew 28 feet. Rocket B flew 19 feet. How much farther did Rocket A fly?	$28 - 19 = 9$ feet
Multiplication	You found 12 rabbits living on 1 acre of land. How many might you find on 5 acres?	$12 \times 5 = 60$ rabbits
Division	A machine moves 600 feet in 1 hour. How many feet does it move in 1 minute?	$600 \div 60 = 10$ feet

Guided Questions

Which two operations can you use to find how many in all?

DIRECTIONS For each question, write your answer in the spaces provided.

1. Why do scientists need good math skills?

   _____

2. What four operations do scientists use?

   _____

3. Solve the problem: A plant grew 2 cm each day. How much did it grow in 1 week? Show your work.

   _____

Math Skills in Science — Lesson 4

Apply the NYS Learning Standards

DIRECTIONS Read the text below and answer the questions.

A scientist recorded some observations. She found four mice that weighed 15 grams each. She fed two mice Diet A. She fed two mice Diet B. Then she weighed the mice after two weeks. Here are the tables she made.

DIET A

	Weight Before	Weight After
Mouse 1	15 grams	21 grams
Mouse 2	15 grams	22 grams

DIET B

	Weight Before	Weight After
Mouse 3	15 grams	17 grams
Mouse 4	15 grams	18 grams

1. How much weight did each mouse gain? Show your work.

   _____
   _____
   _____

2. Which mouse gained the least weight? How much weight did that mouse gain? Which diet did that mouse eat?

   _____
   _____

3. Of the mice that ate Diet A, which one gained more weight? Use less than (<) or greater than (>) signs to show your work.

   _____
   _____

4. Which diet led to the greater gain in weight? How can you tell?

   _____
   _____

Lesson 4 Math Skills in Science

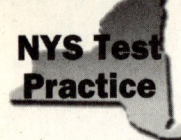 **DIRECTIONS** Choose the best answer for each question. Then circle the letter of the answer you have chosen.

1. Scientists use math skills to
   A   count scientific tools
   B   describe their observations and data
   C   decide what question to ask
   D   number steps in a task

 2. There are 19 tomato plants (T) and 15 pepper plants (P) in a garden. Which number sentence is true?
   A $T = P$
   B $T < P$
   C $P > T$
   D $T > P$

3. There were 16 geese living on a pond all winter. In spring, another 24 geese flew from the south to join them. Which operation would you use to find the number of geese that lived on the pond in spring?
   A   addition
   B   subtraction
   C   multiplication
   D   division

 4. This table shows the height of three trees.

TREE	HEIGHT
Coconut Palm	100 feet
Date Palm	75 feet
Queen Palm	40 feet

Which conclusion could you draw from this information?
   A   The date palm is the shortest of the three trees.
   B   The queen palm is taller than the date palm.
   C   The coconut palm is the tallest of the three trees.
   D   The date palm is taller than the coconut palm.

5. Which statement best describes the pattern in the following numbers?

   1, 3, 5, 7, 9, 11

   A   Each number doubles the one before.
   B   Each number is 2 less than the one before.
   C   The numbers increase by 2, then 3, then 4, and so on.
   D   Each number increases by 2.

Focus on the NYS Learning Standards

Lesson 5: Tools for Experimenting

M3.1a Use appropriate scientific tools, such as metric rulers, spring scale, pan balance, graph paper, thermometers [Fahrenheit and Celcius], graduated cylinder to solve problems about the natural world.

S2.1a Indicate materials to be used and steps to follow to conduct the investigation and describe how data will be recorded (journal, dates and times, etc.)

S2.2a Explain the steps of a plan to others, actively listening to their suggestions for possible modification of the plan, seeking clarification and understanding of the suggestions and modifying the plan where appropriate.

S2.3a Use appropriate "inquiry and process skills" to collect data.

S2.3b Record observations accurately and concisely.

You can use science tools to collect information during an investigation.

A **microscope** is a tool used to observe and measure small objects.

A **meterstick** is a tool used to measure distance and length.

A **thermometer** is a tool used to measure temperature.

A **balance** is a tool used to measure mass.

A **trial** is a completed measurement.

Guided Instruction

DIRECTIONS Read the following information.

You can use tools to help you observe things and collect data. You can use a microscope to see something very small that you normally could not see. You can also use a hand lens to see small objects.

You can use a camera to take photographs. You can use a recorder to collect sounds. You can use a compass to find your direction. You can use a magnet to attract objects made of iron.

There are many tools you can use to measure things. Suppose you want to collect information about how tall you are. You can use a meterstick or a ruler. A **meterstick** measures distance and length.

A **thermometer** measures temperature. In science, temperature is usually measured in degrees Celsius. The symbol for this is: °C.

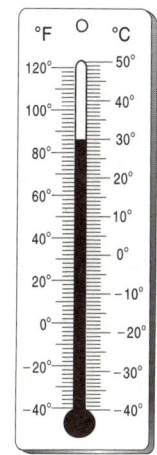

Guided Questions

Why are tools used in science?

What can you use to observe small objects?

What tool would be useful if you were lost?

What is a **meterstick**?

Lesson 5 — Tools for Experimenting

You can use a measuring cup to measure the volume of liquid. For example, you can use a measuring cup to find the volume of a carton of milk.

A balance measures mass. You can use a balance to find out the mass of a rock. Mass is usually measured in grams or kilograms.

During some investigations you will need to measure time. You can use a stopwatch or a clock with a second hand to measure time.

It is a good idea to check your measurements. How do you know if your measurement is a reasonable answer? You can estimate the measurement. Also, you can compare what you are measuring with a similar-sized object. A more accurate way to check a measurement is by doing more than one trial. A **trial** is a completed measurement. For example, students timed how long it took a ball to fall. Then they made this table. Which trial is probably wrong? If one of your measurements is much different from others, you might have made a mistake.

You can use a calculator or a computer to analyze your measurements.

Trial	Time
1	1 second
2	2 seconds
3	10 seconds
4	1 second

Guided Questions

Which tool measures mass?

What is a **trial**?

DIRECTIONS For each question, write your answer in the spaces provided.

1. What is the purpose of doing more than one trial?

   _____

2. What tool could you use to find out how much a small tree has grown? Why would you use this tool?

   _____
   _____
   _____

Tools for Experimenting **Lesson 5**

Apply the NYS Learning Standards

DIRECTIONS Read the text below and answer the questions.

Knowing which tool to use and how to use it will help you collect data. For example, a hand lens is best for small objects you can touch. A microscope is for very small objects that might be harmed if touched. Use the chart to find out how to use tools.

Science Tool	How to Use It
Balance	Put the object you want to measure in the left-hand pan. Put masses you know the weight of in the other pan until the pans are even. Add up the mass in the right-hand pan. That is the mass of the object you measured.
Meterstick	Lay a meterstick beside an object. Match the end of the meterstick to the end of the object. Look at how far the other end of the object stretches along the meterstick. The mark closest on the meterstick is the object's length.

1. Look at the balance. How much mass does the rock have in grams?

2. About what is the length in centimeters of the crayon shown below? About what is the length of the paper clip?

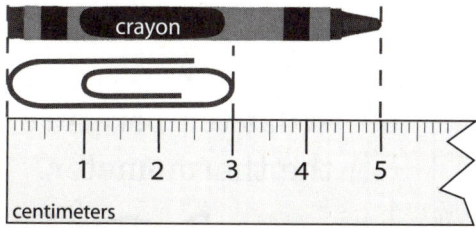

Peoples Education — Copying is illegal. — Chapter 1 • Scientific Inquiry

Lesson 5 Tools for Experimenting

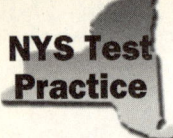

NYS Test Practice

DIRECTIONS Choose the best answer for each question. Then circle the letter of the answer you have chosen.

1 Which of the following tools is used to measure mass?

A

C

B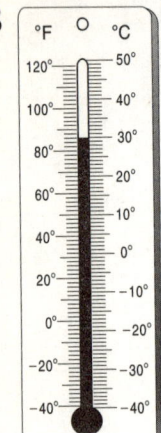

D

2 Data from an investigation included 22°C. What was measured?

A volume
B mass
C temperature
D distance

3 If you wanted to find out the length of your classroom, you could use a

A meterstick
B balance
C calculator
D compass

4 The length of a car is 4 meters. Which of the following is a reasonable measurement of a classroom's length?

A 1 meter
B 9 meters
C 52 meters
D 105 meters

5 You are collecting data that compares the speed of a rabbit with the speed of a dog. Which of the following tools would you use?

A balance and measuring cup
B compass and stopwatch
C stopwatch and meterstick
D meterstick and hand lens

6 What temperature is shown on the thermometer?

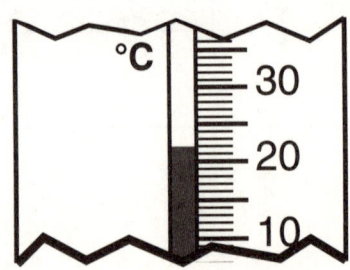

A 14°C
B 20°C
C 22°C
D 34°C

Performance Task

Graph Temperature Data

Focus on the NYS Learning Standards: PS2.1b; S2.1a; S2.2a; S2.3a; S2.3b; S3.1a; S3.2a; S3.3a; S3.4a; ICT6.1; ICT6.5; IPS7.1; IPS7.2

Task:

You will make a bar graph to show average monthly temperatures in Yonkers, New York.

Materials:

- pencil
- red crayon, colored pencil, or marker
- yellow crayon, colored pencil, or marker
- blue crayon, colored pencil, or marker

Procedure:

- Look at the table. Then answer questions 1 and 2.
- Complete a bar graph. Use the information in the table to complete your graph. Use the materials above to complete this task.

Month	Average Temperature in Degrees Celsius
January (Jan)	3
February (Feb)	6
March (Mar)	11
April (Apr)	17
May (May)	22
June (Jun)	27
July (Jul)	30
August (Aug)	28
September (Sep)	24
October (Oct)	18
November (Nov)	12
December (Dec)	6

1. What is the average temperature in Yonkers in May?

   _____

2. What is the average temperature in Yonkers in December?

   _____

- Now, you will complete a bar graph. The graph below shows the average temperatures in different months in Yonkers. Notice that the graph isn't finished. Use the information in the table to finish the graph. Remember to title the graph. Don't color in the bars in the graph yet. Answer the question below the graph.

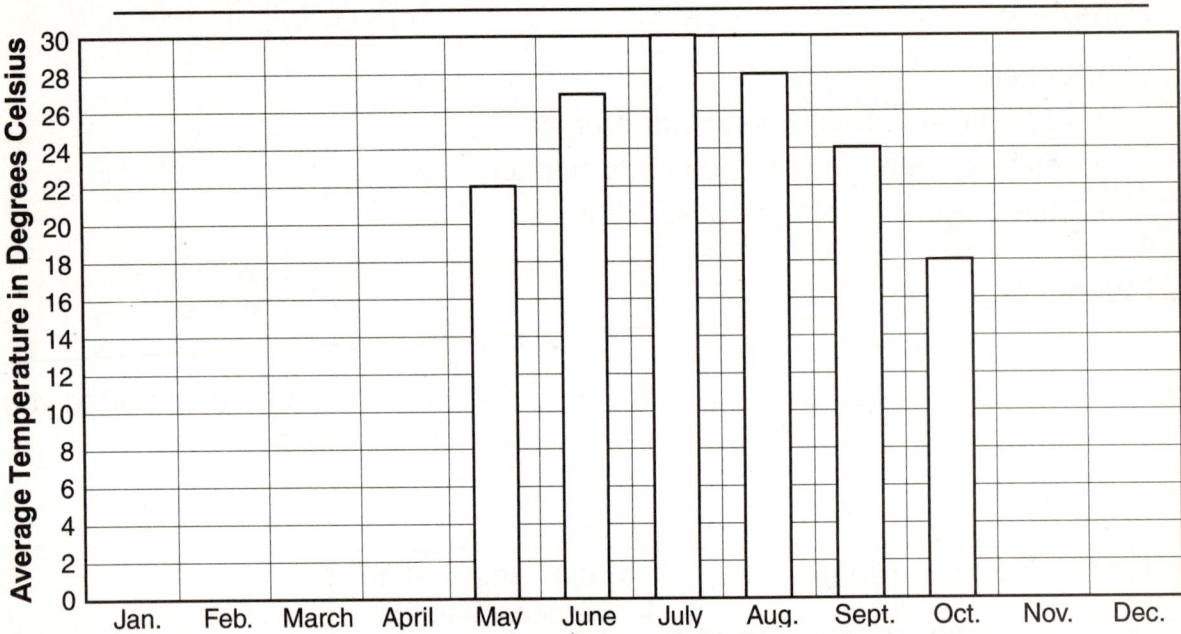

3. Which month, or months, had the highest average temperature?

   _____

4. Now, you will color in the bars in the graph. Find the months that have average temperatures of 17 degrees Celsius or lower. Color the bars for those months blue.

   _____

   _____

5. Find the months that have average temperatures between 18 degrees Celsius and 24 degrees Celsius. Color the bars for those months yellow.

_____

_____

6. Find the months that have average temperatures of 25 degrees Celsius or higher. Color the bars for those months red.

_____

7. How many months have an average temperature of less than 17 degrees Celsius?

_____

8. How many months have an average temperature of between 18 degrees Celsius and 24 degrees Celsius?

_____

9. How many months have an average temperature of greater than 25 degrees Celsius?

_____

10. Which three months are the warmest?

_____

11. Compare the data table and the bar graph that you made. Which one is easier to use to answer question 10? Why?

_____

_____

_____

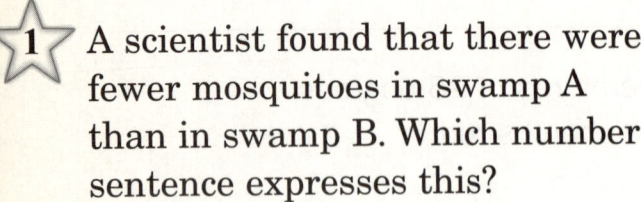

Part I

Directions (1-16): Each question is followed by four choices. Decide which choice is the best answer. Circle the letter of the answer you have chosen.

1. A scientist found that there were fewer mosquitoes in swamp A than in swamp B. Which number sentence expresses this?

    A A = B
    B A < B
    C A > B
    D B < A

2. A scientist found that a new type of bean grew twice as fast as string beans. He concluded that if string beans grew 6 inches in 8 days, the new beans would grow

    A 6 inches in 4 days
    B 8 inches in 6 days
    C 3 inches in 8 days
    D 12 inches in 4 days

3. Which senses could you use to observe a flower?

    A sight, hearing
    B sight, smell, touch
    C smell, taste, touch
    D touch, taste, hearing

4. Suppose a scientist is watching a bird build its nest. Which question might the scientist answer by observing the bird for an hour or two?

    A Why does a bird build a nest?
    B How often does a bird build a nest?
    C What materials does a bird use to build a nest?
    D Do all birds build the same kind of nest?

5. Two students picked up a leaf from the ground. One student found that a leaf measured 5 inches long. Another found that the same leaf measured $5\frac{1}{2}$ inches long. What might cause this difference?

    A They used different tools to measure the leaf.
    B They measured the leaf on different days.
    C One measured in the day, and the other measured at night.
    D They used different hands to hold the ruler.

6 You are asked to group animals according to how they move. In which category would you put this animal?

- A Flies
- B Walks
- C Swims
- D Crawls

7 On Monday, it was sunny all day. There were no clouds in the sky and it did not rain. On Tuesday morning, it was also sunny and there were still no clouds. Which is probably true?

- A It will rain on Tuesday.
- B It will not rain on Tuesday.
- C There will be clouds later in the day.
- D It will not be sunny all day on Tuesday.

8

All of the leaves on a tree have the same shape. The four leaves above

- A did not come from trees
- B came from different parts of the same tree
- C all came from the same tree
- D each came from a different tree

9 What is the length of the leaf to the nearest centimeter?

- A 3 cm
- B 4 cm
- C 5 cm
- D 6 cm

Building Stamina

10. A student wants to know if giving more water to flowers will help them to grow bigger. Which hypothesis could the student use to do an investigation?

   A If I give the flowers more water, then they will grow bigger.

   B If I use different flower seeds, then the flowers will grow bigger.

   C If the flowers get less sunlight, then they will need more water.

   D If the flowers grow better, then they will need less water.

11. Abbie needs to find the mass of a small rock. Which tool should Abbie use?

   A

   B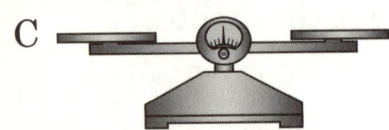

   C

   D

12. During an investigation, you measure the mass of an object. When should you record your measurement?

   A as soon as you make it

   B when you do your homework

   C at the end of the investigation

   D whenever you remember to

13. Maki measures the temperature outside her house every day for a week. What would be the BEST way for Maki to show her class her data?

   A a map

   B a story

   C a graph

   D a picture

14. Jamal records the weather for ten days. On half of the days, the temperature was above 15 degrees Celsius. On how many days was the temperature above 15 degrees Celsius?

   A one   C five

   B two   D ten

15 Jae is writing instructions for an experiment he did. What steps should Jae describe in his instructions?

A   all of the steps he did

B   only the steps that included tools

C   only the steps that included measuring

D   only the steps that included observations

16 Look at the picture below. What is the mass of the rock?

A   50 g

B   150 g

C   250 g

D   350 g

Part II

Directions (17-20): Record your answer on the lines provided below each question.

Students in Syracuse recorded the temperature at noon every day for a week. Here are their findings.

Day of the Week	Temperature in Degrees F
Sunday	40
Monday	37
Tuesday	34
Wednesday	26
Thursday	25
Friday	30
Saturday	34

17 On which day was the noontime temperature coldest?

_____

18 Which two days had the same noontime temperatures?

_____

19 Complete this graph to show the students' findings.

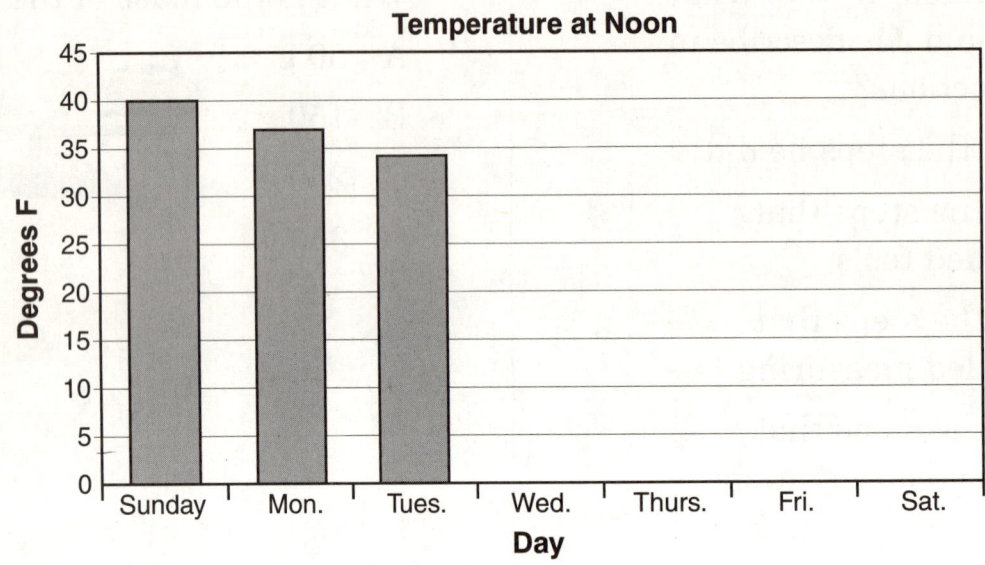

20 In your own words, explain what happened to the temperature over the week.

_____

_____

Focus on the NYS Learning Standards

Chapter 2
Lesson 6 — Properties of Matter

PS3.1a Matter takes up space and has mass. Two objects cannot occupy the same place at the same time.
PS3.1b Matter has properties (color, hardness, odor, sound, taste, etc.) that can be observed through the senses.
PS3.1c Objects have properties that can be observed, described, and/or measured.
PS3.1d Measurements can be made with standard metric units and nonstandard units.
PS3.1e The material(s) an object is made up of determine some specific properties of the object (sink/float, conductivity, magnetism). Properties can be observed or measured with tools such as hand lenses, metric rulers, thermometers, balances, magnets, circuit testers, and graduated cylinders.

You can understand that matter has physical properties.
Mass is the amount of matter in an object.
A **magnet** attracts objects that contain iron.
The five **senses** are sight, hearing, smell, taste, and touch.

DIRECTIONS Read the following information.

Everything around you is matter. **Mass** is the amount of matter in an object. You cannot tell the mass of an object by looking at it. You can find its mass by using a balance. In science, mass is often measured in units called grams.

Balances are used in science to measure and compare the masses of many things. Objects that are the same size can have different masses. Look at a table-tennis ball and a golf ball. Which one has more mass than the other? You can observe this by looking at the balance in the picture. The table-tennis ball is on the left and the golf ball is on the right.

Magnets have a special kind of property. A **magnet** attracts objects that contain iron. You can measure the strength of magnets. Here is one way. Get several magnets and a box of paper clips. See how many paper clips each magnet picks up by putting the

Guided Questions

Which tool measures **mass**?

How do you know which ball has more mass?

What kinds of objects do **magnets** attract?

Lesson 6: Properties of Matter

paper clips in a row touching each other. Place a magnet at the end of the row, and lift the magnet up. The more paper clips a magnet picks up, the stronger the magnet.

Suppose you put gravel and small tacks in a jar and shake it. Whenever you put together two or more kinds of matter and the matter keeps its properties, you create a mixture. Can you separate the gravel and small tacks in the jar by hand? Yes. All you need is a magnet. After the matter in a mixture is separated, it is the same as it was before it was mixed. No new matter is formed.

You can use your **senses** to learn about the properties of matter. You can see an object's color. You can hear its sound. You can scratch it, touch it, or bend it to test its hardness, temperature, or stiffness. You can smell its odor. You can even taste some objects. Your own observations tell you a lot about the properties of matter.

Guided Questions

What do we call two or more kinds of matter that have been combined and can later be separated?

DIRECTIONS For each question, write your answer in the spaces provided.

1. Which of these has more mass: a beach ball or a baseball?

   _____

2. What do you use to find out about the properties of matter?

   _____

3. You place a paper clip on one side of a pan balance. Then you place a pencil on the other side. What is likely to happen? Which is likely to have more mass?

   _____

   _____

Properties of Matter — Lesson 6

DIRECTIONS Study the drawings, read the paragraph, and answer the questions.

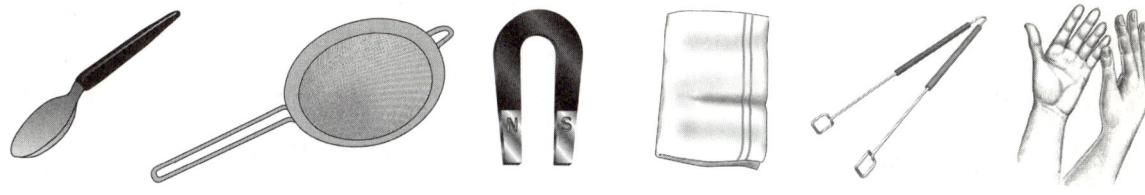

Spoon   Sieve   Magnet   Cloth   Tongs   Hands

Mixtures of matter can be separated, based on their properties. Look at the objects. Think how you can use each item to separate matter. Then complete the chart below.

Separating Matter in a Mixture

What to Do	How to Do It
1. Remove small nails from sand in a sandbox.	
2. Separate the letters in alphabet cereal.	
3. Remove peas from vegetable soup.	
4. Remove seeds from orange juice.	
5. Remove lettuce from a salad.	
6. Remove sand from a pail of water.	

7. Describe three properties of a pencil. Then list the sense you used to describe each property.

_____

_____

_____

Lesson 6 Properties of Matter

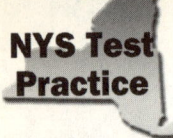 **DIRECTIONS** Choose the best answer for each question. Then circle the letter of the answer you have chosen.

1 Which tool measures mass?
 A beaker
 B thermometer
 C balance
 D magnet

 2 Which can be attracted to a magnet?
 A a plastic button
 B an eraser
 C a marble
 D a paper clip

 3 You can find out how hard an object is by trying to
 A move it with a magnet
 B scratch it or bend it
 C melt it or freeze it
 D lift it or squeeze it

 4 Which sense would you probably not use to observe a fruit tree?
 A smell
 B touch
 C hearing
 D taste

 5 Study the chart below.

Kind of Magnet	Number of Paper Clips
Large bar magnet	25
Small horseshoe magnet	16
Large horseshoe magnet	31
Small bar magnet	22

Which sentence about the information in this chart is true?
 A The bar magnets are the strongest.
 B The horseshoe magnets are the strongest.
 C The largest magnets are the strongest.
 D Size has nothing to do with the strength of magnets.

Focus on the NYS Learning Standards

Lesson 7: Classifying Matter

PS3.1c Objects have properties that can be observed, described, and/or measured.
PS3.1d Measurements can be made with standard metric units and nonstandard units.
PS3.1e The material(s) an object is made up of determine some specific properties of the object (sink/float, conductivity, magnetism). Properties can be observed or measured with tools such as hand lenses, metric rulers, thermometers, balances, magnets, circuit testers, and graduated cylinders.
PS3.1f Objects and/or materials can be sorted or classified according to their properties.
PS3.1g Some properties of an object are dependent on the conditions of the present surroundings in which the object exists.

You can observe and measure objects, and you can classify objects according to their properties.

To **classify** is to group objects based on how they are alike or different. **Properties** are what you can observe about an object. Some properties include size, weight, color, and hardness.

Guided Instruction

DIRECTIONS Read the following information.

When you **classify** objects, you think about how they are alike and different. You look at their **properties**. Look at this sports equipment.

You could classify these objects in many ways. You might put the mitt with the helmet, because you can wear them both. Suppose you classified the objects by shape. You might put the baseball and the basketball together.

When you classify objects, you may think about them in many different ways. You may use your senses to note their shape or hardness. You may use tools to measure their mass or length. You can test their temperatures or see whether they stick to a magnet.

Guided Questions

What properties would you use to **classify** the objects?

What are some tools you might use to classify objects?

Lesson 7: Classifying Matter

The objects here can all be classified as "fruit." You could group the grapefruit, banana, and pear together. They are all yellow. You could group the peach, apples, grapefruit, and orange together. They are all round.

If you use a scientist's eye, you can classify the fruit in different ways. You could put the grapefruit and orange together. They have the same rough peel and are both citrus fruits. You could put the apples and pear together. They have the same kind of stem that attaches them to a tree and similar-looking seeds.

You might use tools to measure the fruit. Maybe the banana and the peach would have the same mass. Maybe the orange and the apple take up the same amount of space. Maybe the peel of the orange and the peel of the grapefruit have the same thickness. Tools can help you classify objects.

Guided Questions

How are the apples and oranges alike? How are they different?

DIRECTIONS For each question, write your answer in the spaces provided.

1. What does it mean to classify objects?

___

2. Name three properties of an orange.

___

3. What tool could you use to measure the length of an object?

___

Classifying Matter **Lesson 7**

DIRECTIONS Read the text below, study the chart, and answer the questions.

Your science class observed some cones on trees around the school. You recorded your observations in a chart.

TREE	OBSERVATIONS OF CONES
White Spruce	Hanging from middle of branch, 2 inches long, smooth
Eastern Hemlock	Hanging from tip of branch, less than 1 inch long
Balsam Fir	Pointing upward, purplish, 3 inches long
Norway Spruce	Hanging downward from middle of branch, brown, about 6 inches long
Tamarack	Small oval cones, less than 1 inch long, pointing up

1. Group the cones by size. Write the name of each tree in the correct column.

LESS THAN 1 INCH LONG	OBSERVATIONS OF CONES

2. Group the cones by type. Write the name of each tree in the correct column.

HANG DOWN FROM BRANCH TIP	HANG DOWN FROM MIDDLE OF BRANCH	POINT UPWARD

Chapter 2 • Matter

Lesson 7 Classifying Matter

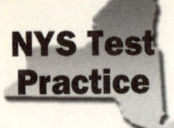

DIRECTIONS Choose the best answer for each question. Then circle the letter of the answer you have chosen.

1. You might use a thermometer to measure
   A. the weight of an object
   B. the width of an object
   C. the volume of an object
   D. the temperature of an object

 2. Look at these shapes.

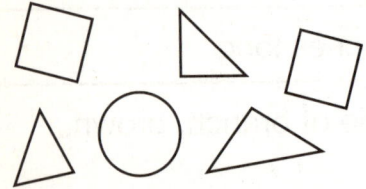

What properties could you use to sort the shapes into two groups of three shapes each?
   A. circles/not circles
   B. squares/not squares
   C. four-sided/three-sided
   D. triangles/not triangles

3. Which property is being measured here?
   A. height
   B. mass
   C. width
   D. shape

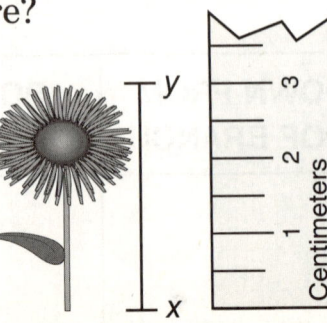

 4. A student classified some classroom objects this way.

GROUP 1	GROUP 2
Paper clip	Pencil
Nail	Book
Stapler	Eraser

Which properties did the student most likely use to classify the objects?
   A. heavier than 1 pound/lighter than 1 pound
   B. colorful/not colorful
   C. attracted to magnet/not attracted to magnet
   D. longer than 1 foot/shorter than 1 foot

5. Which properties best describe an apple?
   A. orange, rough, hard
   B. dull, heavy, cold
   C. soft, green, flat
   D. round, crunchy, smooth

Focus on the NYS Learning Standards

Lesson 8 — States of Matter

PS3.2a Matter exists in three states: solid, liquid, gas.
PS3.2b Temperature can affect the state of matter of a substance.
PS3.2c Changes in the properties or materials of objects can be observed and described.

You can understand the physical properties of matter and temperature.
Matter is anything that has mass and takes up space. Matter has physical properties.
A **physical property** is one that can be observed with the senses.
The three **states of matter** are solid, liquid, and gas.
Volume is the amount of space that matter takes up.
Temperature is the measure of how cold or hot something is.

Guided Instruction

DIRECTIONS Read the following information.

Everything in the world is matter. Matter is anything that has mass and takes up space. Matter has physical properties. A physical property is anything you can observe about an object with your senses. Some things are rough, and others are smooth. Some are sweet, and others are sour. Each word that describes an object names a physical property of the object.

All matter is a solid, a liquid, or a gas. Solids, liquids, and gases are the three states of matter.

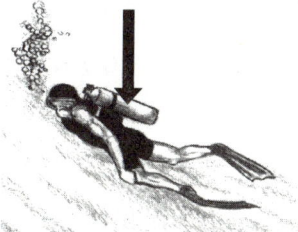

Solid   **Liquid**   **Gas**

Baseballs, rocks, and alphabet blocks are examples of solids. Water and gasoline are liquids. Air is a gas. So is oxygen.

A solid has a definite shape. The shape of a solid does not change. For example, a baseball always has a round shape. It is never square.

Guided Questions

What is **matter**?

How can you tell the **physical properties** of an object?

What are the three **states of matter**?

Lesson 8 — States of Matter

A solid's volume does not change, either. Volume is the amount of space that an object takes up. A baseball takes up the same amount of space day after day.

The volume of a liquid does not change. However, liquids can change shape. For example, if a liquid is poured from a glass into a bowl, its shape changes. It takes the shape of the container.

You can measure the temperature of a liquid. Temperature is a measure of how hot or cold something is. You can use a thermometer to measure temperature.

Like liquids, gases also take the shape of the container they are in. The oxygen in an oxygen tank fills up all of the space in the tank. If the oxygen is moved into a much larger tank, it spreads out and fills all the space in the larger tank. The same amount of oxygen can be packed into a smaller tank, too. So the volume of a gas depends on the size of the container it is in.

Guided Questions

What is **volume**?

What does the volume of gas depend on?

DIRECTIONS For each question, write your answer in the spaces provided.

1. How are gases and liquids different?

   _____

   _____

2. What is the state of matter inside a balloon?

   _____

   _____

3. How is a liquid different from a solid?

   _____

   _____

   _____

States of Matter **Lesson 8**

Apply the NYS Learning Standards

DIRECTIONS Read the paragraph, complete the chart, and answer the question.

A class set up an investigation to find out about changes in physical properties. They listed five plans on a chart. Before the investigation began, the teacher recorded what the students thought would be the result of each plan.

Complete the chart to show what you think would happen, and answer the question below the chart.

Plans	Predictions
1. Place a cup of tap water in a place that is 0°C or less.	
2. Hold a cup of ice cubes in your hands.	
3. Heat a cup of water to 100°C.	
4. Place a cup of tap water in a sunny spot for two days.	
5. Pour cold lemonade into a glass in a place that is about 27°C.	

6. What might you learn about matter from this investigation?

_____

_____

_____

_____

_____

_____

Lesson 8 States of Matter

 **NYS Test Practice**

DIRECTIONS Choose the best answer for each question. Then circle the letter of the answer you have chosen.

1. Oxygen and air are
   A liquids
   B solids
   C gases
   D air

2. Objects have volume. Volume means
   A the amount of space something takes up
   B the weight of air in a balloon
   C how loud something is
   D how heavy something is

3. Which tool measures the temperature of a liquid?
   A tape measure
   B thermometer
   C measuring stick
   D measuring cup

4. What change occurs when an ice cube is heated?
   A It changes from a solid to a gas.
   B It changes from a gas to a liquid.
   C It changes from a liquid to a solid.
   D It changes from a solid to a liquid.

5. Which of the following is an example of gas in a container?

   A

   B

   C

   D

Performance Task

Plan an Experiment

Focus on the NYS Learning Standards: S2.1a; S2.2a; T1.2a; T1.4a; ICT6.3; IPS7.2

Task:

You will write and draw directions to explain to others how to do a science experiment.

Materials:

- pencil
- crayons, markers, or colored pencils

In science, you will carry out experiments. It is important for you to describe exactly what you do during an experiment. That way, someone else can do the same experiment. If you both get the same results, it probably means that your results are correct. This activity will help you learn how to give clear directions.

Procedure:

1. Use the materials above to complete this task.

2. Suppose that you are going to do an experiment to answer this question: "Which can soak up more water: 50 grams of paper towels or 50 grams of cotton balls?" You are not actually going to do this experiment. You are only going to plan and write steps for someone else to follow.

3. First, you need to plan the experiment. What materials do you think you would need for this experiment? You can list the materials you will need below. Or you can draw pictures of all the materials you will need, but don't forget to label them.

_____

_____

4. Now you need to write directions for the experiment. In the space below, describe what you would do to perform the experiment. Number the steps, starting with number 1. Use both words and pictures to describe your experiment.

Building Stamina

Part I

Directions (1–16): Each question is followed by four choices. Decide which choice is the best answer. Circle the letter of the answer you have chosen.

1. Which tool measures the temperature of a liquid?
   A thermometer
   B measuring cup
   C balance
   D ruler

2. Which is an example of a gas?
   A milk
   B water
   C baseball
   D helium

3. Which is a state of matter that keeps the same volume and shape?
   A liquid
   B gas
   C solid
   D air

4. Which of the following is matter you can separate with a magnet?
   A paper clips mixed with rubber bands
   B crayons mixed with pencils
   C pennies mixed with dimes
   D peanuts mixed with popcorn

5. What does gas do that other states of matter do not do?
   A spreads out to fill up containers of different sizes and shapes
   B keeps the same shape when it goes into different containers
   C changes its mass when it goes into different containers
   D always keeps a definite shape, volume, and mass

6. What is the best way to tell the hardness of an object?
   A by trying to dissolve it in water
   B by looking at it
   C by trying to scratch its surface
   D by measuring its mass on a balancing scale

7. Which tool measures mass?

   A

   B

   C

   D

8. Which object will a magnet not attract?

   A

   B

   C

   D

9. Which of the following happens when water becomes a gas?

   A It forms a new kind of matter.
   B It is no longer water.
   C It is gone forever.
   D It changes its state.

10. Which material do magnets attract?

    A glass
    B iron
    C aluminum
    D plastic

11. How do you know that milk is a liquid?

    A Its keeps its shape and volume.
    B It takes the shape of its container and keeps its volume.
    C It keeps its shape but not its volume.
    D It takes the shape and volume of its container.

12. Which properties best describe a rubber band?

    A wide, hard
    B stretchy, flat
    C red, rough
    D magnetic, cold

13 Look at these objects.

What properties could you use to sort the objects into two groups of three objects each?

A used to hit/used to throw

B long/short

C round/not round

D wooden/not wooden

14 How can a very cold temperature affect the state of water?

A It can change a gas to a solid.

B It can change a liquid to a gas.

C It can change a solid to a liquid.

D It can change a liquid to a solid.

15 To classify by length, which unit could you use?

A centimeter

B degree

C cup

D pound

16 Which property of a flower would be affected if it was dark outside?

A its size

B its color

C its odor

D its smoothness

Part II

Directions (17–18): Record your answer on the lines provided below each question.

Imagine that you are asked to do an experiment to answer this question: "Does ice melt faster when it is covered or when it is uncovered?"

17 List the materials you might need to do this experiment.

_____
_____

18 Write the steps of your experiment in order. Number the steps. Then draw a picture showing how the experiment will look.

_____
_____
_____
_____
_____
_____
_____

Chapter 3
Lesson 9 Types of Energy

Focus on the NYS Learning Standards

PS4.1a Energy exists in various forms: heat, electric, sound, chemical, mechanical, light.
PS4.1e Electricity travels in a closed circuit.
PS4.1f Heat can be released in many ways; for example, by burning, rubbing, or combining one substance with another.
PS4.1g Interactions with forms of energy can be either helpful or harmful.

You can identify the many forms of energy.

Energy is the ability to cause change or movement.
Matter is everything around you that takes up space.

Guided Instruction

DIRECTIONS Read the following information.

If you push a door closed, you are using energy. The door moves away from you and it too has energy. Energy makes things happen. It makes things get warm. It makes things move. **Energy** is the ability to cause change or cause movement in matter. **Matter** is everything around you. Energy can change all kinds of matter.

Energy comes in many forms. Some forms include heat, light sound, electric, chemical, and mechanical energy.

For our planet, the Sun is a great source of energy. It provides energy in the form of light. Sunlight is absorbed by the Earth, which changes it to heat energy. The Sun's energy lights and warms the Earth.

Mechanical energy is the energy something has when it is moving. Think about how a school bus works. You probably know that a bus engine uses fuel. Fuel is a kind of chemical energy. Chemical energy is energy that is stored until it is needed. When the fuel is burned, the wheels move. The wheels now have mechanical energy. As the engine runs, it gives off heat energy. When the headlights are turned on, light energy is produced. Even the bus horn produces a kind of energy—sound energy!

Guided Questions

What can **energy** do?

Without **energy** from the Sun, what might the Earth be like?

Lesson 9 Types of Energy

Heat energy can be produced in different ways. Burning fuel releases heat energy. Fuels include gasoline, heating oil, and wood. You can also produce heat energy by rubbing your hands together. In fact, anything that rubs against another thing will produce heat. Try rubbing the eraser end of your pencil back-and-forth on your desk quickly. Then feel the eraser and desk. You made heat. Mixing certain things together also releases heat energy. Scientists often mix acid with water and heat is produced.

Electrical energy can make things happen. Electrical energy is a very useful form of energy. Electricity can make fans turn, bells rings, and lightbulbs light. Electricity travels in a loop of wires. If the loop is broken, the electricity stops. A switch can break the loop when it is turned off. When it is turned on, the electricity starts traveling again.

Guided Questions

What are three ways to release heat energy?

When you switch on a lamp, how would you know that electrical energy is traveling through the loop?

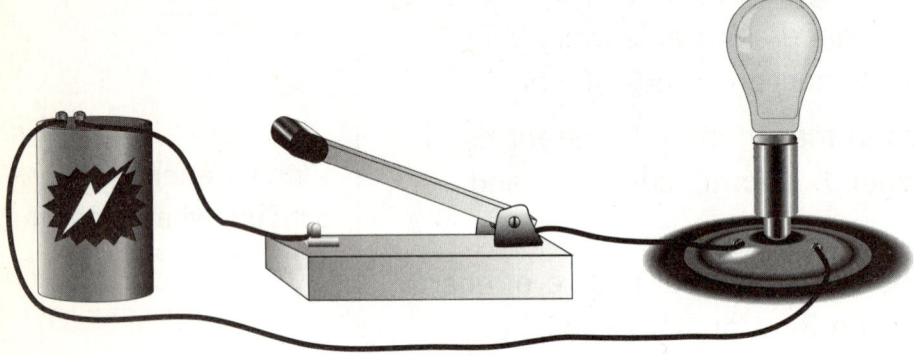

DIRECTIONS For each question, write your answer in the spaces provided.

1. Name five forms of energy that a school bus uses or gives off.

   _____

2. Why might rubbing two sticks together start a fire?

   _____

3. When you switch off the light in your room, what are you really doing?

   _____

Types of Energy **Lesson 9**

Apply the NYS Learning Standards

DIRECTIONS Read the text below and answer the questions.

Energy can be helpful or harmful. Moving air has energy. It moves the leaves on the trees and the grass in the fields. It cools you down on a hot day. It turns windmills, which can be used to produce electricity.

Sometimes, wind energy can be harmful. What if the air moves too fast? It can tear the roof off a house. It can tip over a car.

Other forms of energy can be helpful or harmful too. We love to listen to the sound energy of music. But sounds that are too loud can damage your hearing. And unwanted, continuous sound is called noise pollution. Burning fuels in cars causes air pollution. Too much sunlight can cause sunburn. People need to protect themselves from very hot or very cold temperatures too.

1. What are some ways that humans use the energy from air?

   _____

   _____

2. How might the strong winds of a hurricane be an example of harmful energy?

   _____

   _____

3. Name two machines that you use every day. Tell how they use or give off energy.

   _____

   _____

   _____

4. Name one good effect that comes from burning fuel. Name one bad effect that comes from burning fuel.

   _____

   _____

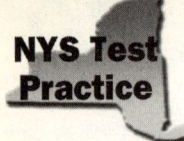

Lesson 9 — Types of Energy

NYS Test Practice

DIRECTIONS Choose the best answer for each question. Then circle the letter of the answer you have chosen.

1. We know that moving air has energy because it can
   - A be burned
   - B move things
   - C give off light
   - D be polluted

2. A whistling teapot might be an example of
   - A electrical energy
   - B mechanical energy
   - C light energy
   - D sound energy

3.  When the wires in the electrical loop on a TV set are broken, what happens?
   - A You see a picture but hear no sound.
   - B You hear sound but see no picture.
   - C The TV set cannot switch on.
   - D The TV set puts out energy.

4. People add oil to a car engine. They do this to keep the parts of the engine from rubbing together too much. What else does oil do to the parts of an engine?
   - A keeps them from getting too hot
   - B keeps them from producing electricity
   - C keeps them from producing light energy
   - D keeps them from producing mechanical energy

5.  Why do we say that we get energy from oil?
   - A We burn oil, producing heat and light.
   - B Oil produces its own chemical energy.
   - C Rubbing oil on our skin produces heat.
   - D Oil can move from one place to another.

6. One example of harmful energy might be
   - A a guitar string
   - B an electric light
   - C a forest fire
   - D a criminal

Focus on the NYS Learning Standards

Lesson 10 — Energy Changing Form

PS4.1b Energy can be transferred from one place to another.
PS4.1c Some materials transfer energy better than others.
PS4.1e Electricity travels in a closed circuit.
PS4.2a Everyday events involve one form of energy being changed to another.
PS4.2b Humans utilize interactions between matter and energy.

You can identify some ways in which energy changes form.

Energy makes things move or change.

A **transfer** of energy is the movement of energy from one place to another. During this movement, energy may change form.

A **circuit** is the closed loop through which electricity travels.

Guided Instruction

DIRECTIONS Read the following information.

Suppose that you have a wagon. The wagon is not moving. Then you pull the wagon. The wagon moves. Where did its **energy** come from? It came from your pull. You **transferred** energy to the wagon.

Guided Questions

Does a wagon move using its own **energy**?

What **transfer** of **energy** takes place when sunlight reaches your skin?

You cannot make energy. You can only change energy that is already there into the kind of energy you need.

Suppose you are going outside on a sunny day. You see the sunlight, but you also feel warm. Some of the light energy from the Sun reaches your skin. It is changed into heat energy. This warms you up.

Lesson 10: Energy Changing Form

Each time you turn on a light, energy is transferred and changes form. Turning on the switch closes a circuit, or a loop. Electricity travels through that loop. It flows through a wire in a bulb. The wire begins to get hot, and then it glows. When you switch on a lamp, electricity changes into heat energy and then into light energy.

Where did the electrical energy come from? It probably came from burning coal. Where did the energy in the coal come from? It came from the energy in plants. Where did the energy in the plants come from? It came from the Sun. The Sun's light energy is very important to us here on Earth.

Green plants capture energy from the Sun. They use that energy to grow. They store some of that energy in their stems and leaves.

When we eat green plants, we change their stored energy into heat and motion. The act of eating is one kind of transfer of energy.

When we chop down a tree and burn the wood, we change its stored energy into heat and light. Burning fuel is another kind of transfer of energy.

Why do we use metal pots to cook our food? Metal transfers heat energy well. Metal pots heat food quickly. Why do we coat electrical wires with plastic? Plastic does not transfer electrical energy well. Plastic coating protects us from electricity.

Guided Questions

If a light is off, is the electrical **circuit** open or closed?

What two **transfers** of **energy** take place in a light bulb?

What are two ways we use stored energy from plants?

What might happen if we did not coat electrical wires with plastic?

DIRECTIONS For each question, write your answer in the spaces provided.

1. What kind of energy changes into light energy in a lightbulb?

2. How does eating transfer energy?

3. Why is metal a good choice for making a frying pan?

Energy Changing Form — **Lesson 10**

Apply the NYS Learning Standards

DIRECTIONS Read the text below and answer the questions.

Many everyday items transfer energy from one form to another. When someone calls on the phone, electrical energy transfers to sound energy, and you hear the ring. When you turn on a flashlight, chemical energy in the battery changes to electrical energy. The bulb lights up.

Why does the grocery store door opens as you walk toward it? The door has an electric eye. Light hits the eye. The metal in the eye produces electricity. The electricity flows in a closed circuit. As you walk toward the door, you block the light. The circuit is broken. Mechanical energy in a motor opens the door.

Sometimes we get electrical energy from burning fuels, such as coal. Sometimes we get electrical energy from moving water. Energy from the Sun and wind can also be transferred into electrical energy. Understanding the transfer of energy is important. It helps scientists find new and better ways of making things work.

1. Name an object that transfers electrical energy to sound energy.

   _____

2. What happens when you turn on a flashlight?

   _____

   _____

3. Some garage doors have electric eyes. If you try to get into the garage when the door is coming down, it goes back up. Explain how this happens.

   _____

   _____

   _____

4. What are four possible sources of electrical energy?

   _____

Lesson 10 — Energy Changing Form

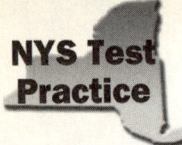 **DIRECTIONS** Choose the best answer for each question. Then circle the letter of the answer you have chosen.

1. Throwing a ball is an example of
   A sound energy
   B energy transfer
   C electricity
   D burning fuel

2. Electricity can flow as long as
   A a circuit is broken
   B the light is on
   C a circuit is closed
   D the switch is off

 3. Which would probably be best at transferring electrical energy?
   A plastic tubes
   B rubber mats
   C metal wire
   D plant stems

4. How does the Sun's energy affect the ocean?
   A It heats it.
   B It adds salt.
   C It makes waves.
   D It makes sound.

 5. A doorbell is an example of a transfer from
   A electrical energy to mechanical energy
   B sound energy to electrical energy
   C light energy to sound energy
   D electrical energy to sound energy

6. Where do plants get the energy that they store?
   A from water
   B from the Sun
   C from the Earth
   D from their leaves

Focus on the NYS Learning Standards

Lesson 11 Heat on the Move

PS4.1b Energy can be transferred from one place to another.
PS4.1c Some materials transfer energy better than others.
PS4.1d Energy and matter interact.
PS4.1f Heat can be released in many ways; for example, by burning, rubbing, or combining one substance with another.

You can learn how heat is made and how it moves between objects.
The **temperature** of something is how hot or cold it is.
To **evaporate** is to change from liquid to gas.

Guided Instruction

DIRECTIONS Read the following information.

Suppose you are having soup for lunch. You hold the bowl in your hands. When you put the bowl down, your hands feel warm. Why are your hands warm? Heat moved from the bowl of soup to your hands.

When heat moves between two objects, it always moves from the warmer object to the colder one. It never moves from the colder object to the warmer one. The bowl of soup is hotter than your hands. When you touch the bowl, the heat moves from the bowl to your hands. If you hold an ice cube in your hand, heat moves from your hand to the ice cube. The heat from your hand makes the ice cube melt. The cold does not move from the ice cube to your hand, even though it may feel that way!

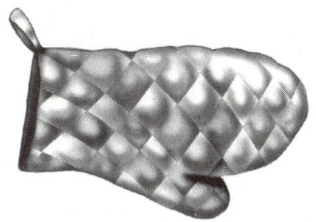 

You cook your soup in a metal pot because metal moves heat energy well. You pick up the pot with an oven mitt. The material in the mitt blocks some of the heat energy. It does not move heat energy well. The mitt protects your hand.

How could you learn how hot your soup was? Remember that you can use a thermometer to measure the **temperature** of something. The temperature tells you how hot or cold something is.

Guided Questions

In which direction does heat always move between a hot object and a cold object?

How is heat moving to melt the ice cubes?

What does a thermometer tell you?

Lesson 11 Heat on the Move

Where did the heat in the soup come from? It may have come from a stove. Stoves that burn gas are sometimes used to cook food. The burning gas makes flames that heat the food. You can also burn wood to make heat. If you stand near a campfire, you will feel the heat of the burning wood. Some people use heat from burning wood to keep their homes warm. They may have a fireplace or a wood-burning stove.

Another way to make heat is by rubbing two things together. Try rubbing your hands together quickly. What happens? Your hands get warmer. The more you rub them together, the warmer they get.

Another way to make heat is by mixing things together. Some chemicals make heat when they are mixed together. For example, some people use small packets called hand warmers to keep their hands warms in the winter. Hand warmers are made of special chemicals. When the chemicals inside the packet mix together, they make heat.

Heat energy can change things. A puddle of water seems to dry up. Where does the water go? Light energy from the Sun becomes heat energy. It heats the water. The water **evaporates**. The water is now a gas, and it mixes with the other gases in the air.

So heat can change liquid to gas. It can also change solid to liquid. What happens when you melt an ice cube by heating a pan? It turns from a solid ice cube into liquid water.

Guided Questions

What happens if you rub two things together?

What happens to water that **evaporates**?

DIRECTIONS For each question, write your answer in the spaces provided.

1. Why does an ice cube melt when you hold it in your hands?

_____

_____

2. What are three ways to make heat?

_____

3. What is evaporation?

Heat on the Move   **Lesson 11**

Apply the NYS Learning Standards

DIRECTIONS Study the drawing and answer the questions.

ice water   iced tea

1. What was the temperature of the ice water at 10:00? What was the temperature of the tea at 10:00?

   _____

   _____

2. What was the temperature of the water at 12:00? What was the temperature of the tea at 12:00?

   _____

   _____

3. Why did the hot tea become cooler over time?

   _____

   _____

   _____

4. Why did the ice water become warmer over time?

   _____

   _____

   _____

Lesson 11 Heat on the Move

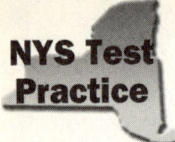 **DIRECTIONS** Choose the best answer for each question. Then circle the letter of the answer you have chosen.

1. Heat always moves from
   A cold to warm
   B top to bottom
   C warm to cold
   D ice to water

2. We use metal pans to cook because
   A they stay cool
   B they burn easily
   C they are always hot
   D they transfer heat well

3. An example of evaporation might be
   A a pan of water turning to ice
   B some spilled water "drying up"
   C a stream of water flowing downhill
   D a pool of water filling up

4. Rosa and Carlos are on a camping trip. They cook their food over a campfire. The heat that cooks their food is made by
   A shaking
   B rubbing
   C mixing
   D burning

5. The brakes of a car rub against the wheels to slow the car down. This rubbing makes the brakes
   A wet
   B hard
   C quiet
   D hot

Lesson 12 Energy and Matter

Focus on the NYS Learning Standards

PS4.1c Some materials transfer energy better than others.
PS4.1d Energy and matter interact.
PS4.2a Everyday events involve one form of energy being changed to another.
PS4.2b Humans utilize interactions between matter and energy.

You can identify ways that energy makes things happen.
Energy makes things move or change.
Solar energy is energy from the Sun.
Matter is anything that has mass and takes up space.

Guided Instruction

DIRECTIONS Read the following information.

The Sun's **energy** makes things on Earth warm. We call energy from the sun **solar energy**. When sunlight shines on an object, the object soaks up the sunlight. The light energy turns into heat. Heat makes the object feel warm. If you sit outside on a warm, sunny day, you may feel heat on your skin and clothes. If you sit in the shade, you will feel less heat because less sunlight falls on you.

During the day, there is plenty of sunlight to make heat. Even on a cold day, sunlight can turn into heat energy. However, when it is night, there is no sunlight to shine on things. Objects cannot soak up sunlight and get warmer. This is why days are warmer than nights.

Some objects can soak up more sunlight than others. How much sunlight an object soaks up depends on two things: what the object is made of and what color it is. The more sunlight something soaks up, the hotter it gets. Metal can soak up a lot of light. For example, a metal plate left out on a sunny day may feel hot if you touch it.

A metal pie plate gets warm when the Sun shines on it.

Sand on a beach or in a sandbox may also get very hot. Things made of wood or plastic do not get as hot as metal or sand. The plastic lights and glass

Guided Questions

How does the Sun's **energy** heat things on Earth?

What two things control how hot something will get in the Sun?

Chapter 3 • Energy 59

Lesson 12 Energy and Matter

windows on a car do not get as hot as the metal parts of the car. Trees, grass, water, and cloth are some other things that do not get as hot as sand or metal.

How warm something gets also depends on what color it is. Darker colors can usually soak up more light than lighter colors. Paved streets may get very hot in sunlight if they are black. The white lines on the street may not get as hot because they are light-colored. The metal in a dark blue car may get hotter than the metal in a tan car, even though they are both metal.

We use solar energy to keep warm. We use it in other ways, too. Solar panels on a rooftop collect solar energy to make hot water for the building. Some calculators use solar energy. They don't need batteries. Just keep them in the sunlight. These calculators change sunlight into electricity.

As energy changes **matter**, it often makes something we can use. We can use mechanical energy to change the sound energy produced by a trumpet. That helps us make music. We can use the stored chemical energy in a candle to make light energy. That helps us see in the dark. We can use the stored chemical energy in oil to make heat energy. We burn oil to produce heat energy. That helps keep us warm, even when the Sun is not out.

Guided Questions

How does color control how hot something gets in the Sun?

How does a solar calculator work?

How do we use the stored chemical **energy** in oil?

DIRECTIONS For each question, write your answer in the spaces provided.

1. How does sunlight make Earth warm?

   _____

2. Does sunlight warm up all things equally? Explain your answer.

   _____

3. Why is nighttime colder than daytime?

   _____

4. How do we use solar energy?

   _____

Energy and Matter **Lesson 12**

Apply the NYS Learning Standards

DIRECTIONS Read the following information and answer the questions.

Amy does an experiment to see how different objects get warm in sunlight. She puts a small table outside in the sunshine. Then, she puts the following items on the table: a wooden log, a black rock, a white rock, a metal toy car, and a plastic toy boat. She waits 1 hour. Then, she touches the items one at a time to see how warm they feel.

1. Which object will probably be warmer: the black rock or the white rock? Explain your answer.

   _____
   _____
   _____

2. Amy observes that the toy car is warmer than the toy boat. Why is the toy car warmer?

   _____
   _____
   _____

3. Amy feels the top side of the log and the bottom side of the log. Which side is probably cooler? Why?

   _____
   _____

4. If Amy covers the objects with a cloth, how will this change how hot they get? Why?

   _____
   _____

Lesson 12 — Energy and Matter

NYS Test Practice

DIRECTIONS Choose the best answer for each question. Then circle the letter of the answer you have chosen.

1 Which of these statements about how sunlight warms things is false?

   A Some colors heat up more than other colors.

   B Some shapes heat up more than other shapes.

   C Objects made of different materials heat up differently.

   D Objects get hot in the Sun because they soak up light.

2 Which of these will heat up the most in the Sun?

   A metal
   B paper
   C plastic
   D wood

3 What happens in a solar calculator?

   A Energy changes to matter.
   B Sunlight changes to electricity.
   C Heat changes to light.
   D Chemical energy is stored.

4 Sunlight makes Earth warm because

   A sunlight is very bright, which makes it hot

   B sunlight turns to heat when things soak it up

   C sunlight keeps heat from moving away from Earth

   D sunlight traps heat from the Sun and brings it to Earth

5 Which shows a way that people use stored chemical energy?

   A burning a candle to give light
   B blowing air through a trumpet
   C working on a solar calculator
   D drinking cool water on a hot day

Focus on the NYS Learning Standards

Lesson 13 Sound Energy

PS4.1a Energy exists in various forms: heat, electric, sound, chemical, mechanical, light.
PS4.1b Energy can be transferred from one place to another.
PS4.1c Some materials transfer energy better than others.
PS4.1d Energy and matter interact.
PS4.1g Interactions with forms of energy can be either helpful or harmful.

You can explore the movement of sound energy.

Sound energy is mechanical energy that causes sound.
To **vibrate** is to move back and forth.
Sound waves are vibrations in air or other matter, started by a vibrating object.
Noise pollution is sound that is too loud or lasts too long and may be harmful.

Guided Instruction

DIRECTIONS Read the following information.

Your friend plucks a string on a guitar. You hear it with your ears. What moves between the guitar and your ears? You can't see it or feel it. It is sound energy.

When the string is plucked it begins to vibrate. It moves back and forth quickly. That vibration makes the air vibrate. The vibrations push through the air in sound waves. The waves move until they reach your ear.

Guided Questions

What causes the guitar string to **vibrate**?

How does that vibration reach your ear?

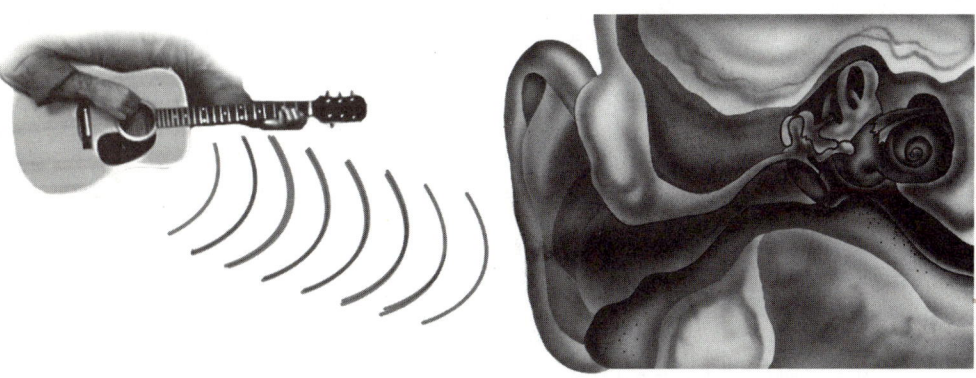

Sound moves quickly through gases, such as the air. It moves a little faster through liquids. But sounds waves move the fastest through solids, such as metal, wood or glass. Very gently tap the tip of your pencil on your desk. Make sure you tap it so you can barely hear it. Now put your ear on your desk and continue tapping gently. Are your surprised that you can hear the tapping better? That is because sound waves travel faster and better through solid materials.

Does sound move fastest through solids, liquids, or gases?

Lesson 13 Sound Energy

Sound energy is important. You listen to music. You pay attention in class. You hear sounds that protect you—a police whistle or a smoke alarm. All of these sounds are made by something that is vibrating. All of these sounds travel in waves to your ears.

Not all sound energy is helpful. **Noise pollution** is sound energy that can be harmful. It can be too loud and go on too long. It can damage your hearing. It can keep you from sleeping. Be careful to protect your ears from noise pollution. Do not play near loud machines. Do not listen to very loud music.

Guided Questions

What are two ways you use sound energy?

Why should you protect your ears from **noise pollution**?

DIRECTIONS For each question, write your answer in the spaces provided.

1. What moves between a ringing telephone and your ears?

   _____
   _____
   _____

2. What does a guitar string do when it vibrates?

   _____
   _____
   _____

3. Why shouldn't you sit right next to a noisy machine?

   _____
   _____
   _____

Sound Energy — Lesson 13

Apply the NYS Learning Standards

DIRECTIONS Read the text below and answer the questions.

A loud sound uses more energy than a soft sound. You can tell that this is true if you tap your pencil on your desk. It takes more energy to make a loud sound. You have to tap harder. Tapping harder makes the desk vibrate more strongly. The strong vibrations travel farther. The loud sound can be heard farther away.

A high sound comes from faster vibrations than a low sound. You can tell that this is true if you pluck a rubber band. If the band is not moving fast, the sound will be lower. If you pull the band tight and pluck it, it will move much faster. The sound will be higher.

Hold your hand against your throat. Hum a low note. You can feel the vibrations that help to make that sound. The vibrations are slow. Now hum a high note. The vibrations are faster. The note sounds higher.

1. Which sound uses more energy, the chirp of a bird or the bark of a dog?

2. Why can you hear a loud sound even if you are far away?

3. If you tighten the string on a guitar, will the sound it makes be higher or lower than before?

4. A cello makes a low sound. A violin makes a high sound. Which one's strings vibrate faster?

Lesson 13 Sound Energy

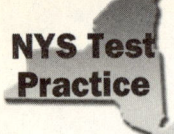

NYS Test Practice

DIRECTIONS Choose the best answer for each question. Then circle the letter of the answer you have chosen.

1. Which is true?
   A Sound energy cannot be made underwater.
   B Sound energy moves fastest in the open air.
   C Sound energy can be seen as well as heard.
   D Sound energy is made by something that vibrates.

2. As a guitar string vibrates in air, it makes
   A heat
   B waves
   C wind
   D pollution

3. In which substance do sound waves travel the fastest?
   A salt water
   B glass
   C fresh water
   D air

4. Which one would make a high sound?
   A a slow-moving fan
   B a ticking watch underwater
   C an insect's fast wings
   D a book dropped on the floor

5. Which is most likely to be a source of noise pollution?

   A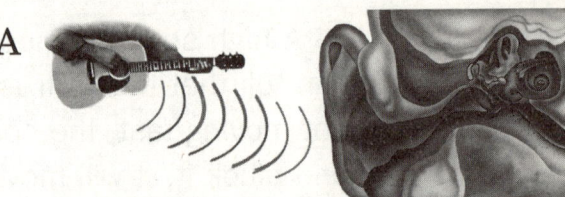

   B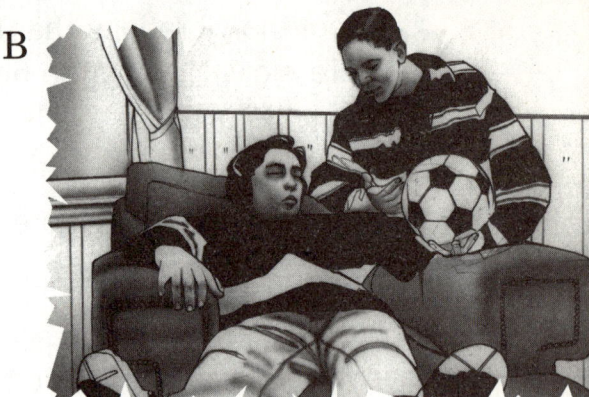

   C

   D

Performance Task

Build Insulated Containers

Focus on the NYS Learning Standards: PS4.1b,c; M3.1a; T1.2a,b; T1.3a,b,c; T1.4a,b; T1.5b,c; ICT6.2; IPS7.1; IPS7.2

Task:

Insulation is a material that keeps heat from moving. Remember that heat always moves from hotter things to colder things. Many things can be insulation: fur, hair, feathers, fat, cloth, and even air. Some materials are better than others at keeping heat from moving. Therefore, some materials are better for insulation than others. You will make two containers out of different materials. Then you will find out which one keeps water warmer.

Materials:

- pencil
- thermometer
- measuring cup
- hot water
- tape
- glue
- other materials supplied by your teacher

Directions:

1. In this task, you will choose and insulate two containers out of different materials. Then, you will find out which container keeps water warmer.

2. First, look at the materials your teacher gave you. Think about how to make your containers from some of those materials.

3. Decide which materials you will use to make each of your containers. Fill in the table below. You do not have to use all the materials your teacher gave you.

	Container 1	Container 2
Materials I will use for this container		

4. In the space below, describe how you will make each container out of the materials you listed in the table. List two reasons why you think this container will keep water warm.

_____
_____
_____
_____
_____

5. Now, make your containers and put them on a desk or counter. Make sure they are not near anything that could tip them over.

6. Use the measuring cup to measure $\frac{1}{2}$ cup of warm water. Pour the water into Container 1. Repeat this step for Container 2.

7. Use the thermometer to measure the temperature of the water in each container. Write the temperatures you measure in the column labeled "Starting Temperature" in the table below.

Container	Starting Temperature	Ending Temperature	Temperature Change
1			
2			

8. Let your containers sit on the desk or counter for 45 minutes or until your teacher tells you to finish up your experiment. Be careful not to touch the containers or spill the water in them. While you wait, think about what you expect to happen. Which container do you think will keep water warmer? Why?

_____
_____

9. After 45 minutes, use the thermometer to measure the temperature of the water in each container again. Record your measurements in the column labeled "Ending Temperature" in the table in step 7.

10. Next, figure out how much the water in each container has cooled. To do this, subtract the ending temperature for each container from the starting temperature.

Starting temperature − ending temperature = temperature change

Container 1: _____ − _____ = _____

Container 2: _____ − _____ = _____

11. Write the temperature change for each container in the column labeled "Temperature Change" in the table in step 7.

12. Which container had the greater temperature change?

_____

_____

13. Which container was the better insulator?

_____

_____

14. Compare your results with the results of your classmates. Which type of container was the best insulator? Which was the worst?

_____

_____

_____

_____

15. If you wanted to keep hot chocolate warm on a cold day, which insulator would you choose? Explain your answer.

_____

_____

_____

Part I

Directions (1–16): Each question is followed by four choices. Decide which choice is the best answer. Circle the letter of the answer you have chosen.

1. Which object changes mechanical energy to sound energy?
   A a flashlight
   B a candle
   C a birdfeeder
   D a music box

2. A bedside lamp produces
   A light energy
   B chemical energy
   C electrical energy
   D mechanical energy

3. How does the Sun's energy affect a rooftop?
   A It moves it.
   B It sets it on fire.
   C It warms it.
   D It evaporates it.

4. You use a pot holder to pick up a pan because it
   A keeps the pan from moving
   B blocks some of the movement of heat
   C closes an electrical circuit
   D lowers the pan's temperature so you don't burn your hands

5. If an electrical circuit is closed,
   A electricity can flow through it
   B it lights up
   C electricity is blocked
   D it stops making a sound

6. How can sound energy be helpful?
   A It can hurt your ears.
   B It can light your room.
   C It can warn you.
   D It can keep you from sleeping.

7 Violet puts a container of water in a hot room. The room is hotter than the water in the container. The water in the container will most likely

A get cooler
B get warmer
C break the container
D spill out of the container

8 Mara walks in a crowded parking lot on a sunny day. She notices that the parking lot is very hot. Where did the heat most likely come from?

A the Sun
B her body
C the cars on the parking lot
D insulation in the parking lot

9 Frankie's hands are cold. He wants to warm them up. Which of these would be the best way for Frankie to warm his hands?

A Rub them together.
B Hold them far apart.
C Wave them in the air.
D Clap them together once.

10 What are three ways to make heat?

A burning, rubbing, mixing
B cooking, rubbing, boiling
C burning, mixing, folding
D cooking, mixing, tearing

11 Which of these is most likely to make heat?

A water inside a pipe
B a pair of winter gloves
C a knife slicing through an apple
D brakes rubbing on a bicycle tire

Building Stamina

12 Ramon puts ice cubes into a cup of water. After a while, the water is colder and the ice cubes are melted. Which of these best explains why the water gets colder and the ice cubes melt?

 A Heat moves from the water into the ice cubes.

 B Heat moves from the ice cubes into the water.

 C Cold moves from the water into the ice cubes.

 D Cold moves from the ice cubes into the water.

13 A solar battery gets its energy from

 A electricity
 B a car
 C heat
 D the Sun

14 As a violin string vibrates in air, it produces

 A waves
 B heat
 C light energy
 D electricity

15 Which one transfers sound energy best?

 A water
 B oil
 C metal
 D air

16 Which sound uses more energy?

 A

 B

 C

 D

Part II

Directions (17–20): Record your answer on the lines provided below each question.

A scientist developed a new kind of material. He wanted to see how well the material worked as an insulator. He filled three glass jars with hot water. He wrapped Jar A in his new material. He wrapped Jar B in cotton. He left Jar C alone. Here are his results.

TEMPERATURE OF WATER

	Start	After 1 hour	After 2 hours	After 3 hours
Jar A	200°F	185°F	175°F	165°F
Jar B	200°F	150°F	100°F	80°F
Jar C	200°F	120°F	90°F	68°F

17 What two measuring tools did the scientist need to perform his experiment?

_____

18 Should all three jars start out at the same temperature? Why or why not?

_____

19 Draw a picture of the scientist's experiment.

20 Based on the findings, would you say that the scientist's material is a good insulator? Explain your answer.

_____
_____

Focus on the NYS Learning Standards

Chapter 4
Lesson 14 — Forces and Motion

PS5.1a The position of an object can be described by locating it relative to another object or the background.
PS5.1b The position or direction of motion of an object can be changed by pushing or pulling.
PS5.1d The amount of change in the motion of an object is affected by friction.

You can understand how forces change the motion of objects.

Position is where an object is in relation to other objects or the background.

Motion is a change of position.

A **force** is a push or pull on an object that causes an object's position to change.

Friction is a force that stops an object from moving or slows the motion of a moving object.

Guided Instruction

DIRECTIONS Read the following information.

Look around. You see different objects. How would you describe where the objects are? You might say that two objects are next to each other, or that one was on top of another. Words such as *next to*, *on top of*, *over*, and *under* tell about an object's **position**.

What can you see that is moving? You see things moving wherever you go. You may see that some things move a lot and others move only a little. Mountains and buildings do not seem to move at all. Some things, such as Earth, air, and the ocean, are always in motion. **Motion** is a change of position.

Motion does not just happen. Something causes it. That something is a force. A **force** is a push or a pull. It takes a push to start a skateboard moving. It takes a pull to drag a wagon up to the top of a hill. Each push or pull that makes an object move is a force.

Moving water can be a force. It turns waterwheels, washes off sidewalks, and sometimes causes floods. Moving water carries rocks and soil from streams into rivers, and then into oceans.

Guided Questions

What is **motion**?

What causes **motion**?

Moving water is a force.

Moving air, called wind, can also be a force. It causes pinwheels to spin and hats to blow off. It can cause dangerous storms, such as hurricanes and tornadoes, too.

Moving air makes your kite rise in the sky and dance about. When the wind dies down, you know your kite-flying is about to end.

Another force, called **friction**, is at work when you skateboard. There is friction whenever an object starts moving or is moving or stops moving. When you skate, there is friction between the wheels and the surface you are skating on. Friction slows or stops the motion of objects. Rough surfaces cause more friction than smooth surfaces.

Guided Questions

What does **friction** do?

Moving air is a force.

DIRECTIONS For each question, write your answer in the spaces provided.

1. What causes the force that makes kites rise high in the sky?

   _____
   _____

2. What are some words that describe an object's position?

   _____
   _____

3. You give a wagon a push on a sidewalk. It moves, and then it stops. What force helped stop the wagon?

   _____

Lesson 14 Forces and Motion

Apply the NYS Learning Standards

DIRECTIONS Read the paragraph and the steps below, and answer the questions.

You plan to test different surfaces to find the answer to the question, "Which surface causes the least friction?"

Steps in the Investigation

1. Choose three surfaces to test.
2. Make a ramp like the one in the picture.
3. Place the ramp on one of the surfaces.
4. Set a toy car at the top of the ramp, and let it roll down.
5. Measure the distance that the car travels before it stops, and record the distance on a chart.
6. Repeat Step 3 through Step 5 of the test on two other surfaces.

1. Which three surfaces would you test in the investigation?

_____

_____

2. What do you predict you will find out from this investigation? Give a reason for your prediction.

_____

_____

_____

3. Which two things touch each other in the investigation?

_____

4. How can you tell which surface causes less friction?

_____

_____

Forces and Motion **Lesson 14**

NYS Test Practice

DIRECTIONS Choose the best answer for each question. Then circle the letter of the answer you have chosen.

1. A push or a pull on an object is called a

    A force

    B teeter-totter

    C tug-of-war

    D position

2. Which of the following is not a force?

    A the push of moving air

    B the push of moving water

    C clouds in the sky

    D friction between objects

 3. The main force that causes a tornado is due to which of the following?

    A position

    B friction

    C moving air

    D moving water

 4. Which force keeps a school bus from skidding on a curve?

    A moving water on the road

    B the force of the bus driver's hands on the wheel

    C moving air that pushes against the bus

    D friction between the wheels and the road

5. How would you describe the position of the turtle?

    A under the rock

    B over the plant

    C on top of the container

    D next to the water

Peoples Education   Copying is illegal.   Chapter 4 • Forces and Machines   **77**

Focus on the NYS Learning Standards

Lesson 15: Levers, Pulleys, and Inclined Planes

PS5.1f Mechanical energy may cause change in motion through the application of force and through the use of simple machines such as pulleys, levers, and inclined planes.

You will learn how levers, pulleys, and inclined planes make work easier.
A **force** is a push or a pull.

Guided Instruction

DIRECTIONS Read the following information.

When you use force to move an object, you are doing work. A force is a push or a pull. Simple machines, such as levers, pulleys, and inclined planes, change the force used to do work.

Lever

A bar that moves on top of a fixed point is a lever. The fixed point is called a fulcrum. The object that is lifted is called the load. When you push down on one side of the lever, you use force. The lever turns on the fulcrum and lifts the load on the other side.

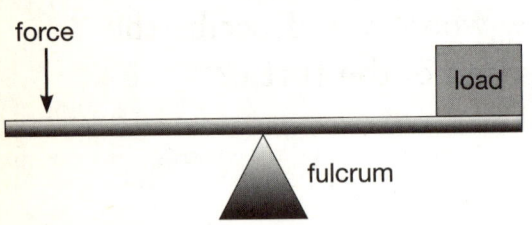

A lever makes work easier by changing the amount of force needed. It takes less force to push down on the lever than to lift the load with your hands. A seesaw is a lever that lets you easily lift a person.

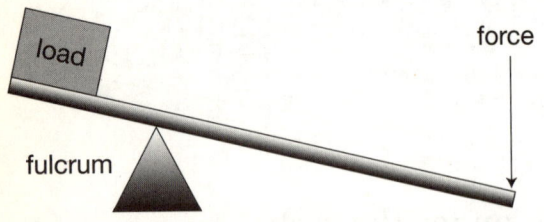

What if you move the fulcrum of the lever closer to the load you are lifting? It will take even less effort to lift the load. The closer the fulcrum is to the load, the less force is needed to lift the load.

A bottle opener is another type of lever. The load is the bottle cap.

Guided Questions

What is the fixed point at the center of the **lever** called?

How does a **lever** change the **force** needed to do work?

Pulley

A pulley is made up of a rope or chain wrapped around a wheel. A pulley can be fixed or movable.

In a fixed pulley, the pulley stays in one place. A load is attached to one end of the rope. When you pull down on the other end of the rope, the wheel turns and the load is lifted. A fixed pulley changes the direction of the force. Pulling down on the rope or chain is easier than lifting up the load by hand.

In a movable pulley, the pulley moves up and down with the load. One end of the rope is attached to a fixed point. You pull the other end up to lift the load.

A moveable pulley does not change the direction of the force. Less force is needed to lift the load.

Inclined Plane

Suppose you need to move a wheelbarrow up some steps. You can put a board at a slant from the bottom to the top. Then you can wheel the wheelbarrow up the steps more easily. The board helps you do work. This slanted board is one kind of inclined plane.

An inclined plane changes the size of the force you must use to do work. It makes the work easier by making the needed force less.

Levers, pulleys, and inclined planes are simple machines. Most of the helpful tools you use are based on simple machines. Simple machines change the force needed to do work.

Guided Questions

What are the two types of **pulleys**?

What do you do with the two ends of rope for a moveable pulley?

How does an **inclined plane** help you do work?

Lesson 15 — Levers, Pulleys, and Inclined Planes

DIRECTIONS For each question, write your answer in the spaces provided.

1. How is a seesaw like a lever?

   _____

   _____

2. What is a fulcrum?

   _____

3. What happens when you move the load on a lever closer to the fulcrum?

   _____

4. How does a fixed pulley make work easier?

   _____

5. How does an inclined plane make work easier?

   _____

   _____

Levers, Pulleys, and Inclined Planes **Lesson 15**

DIRECTIONS Study the drawings and answer the questions.

Brendon is experimenting with levers. He sets up these levers.

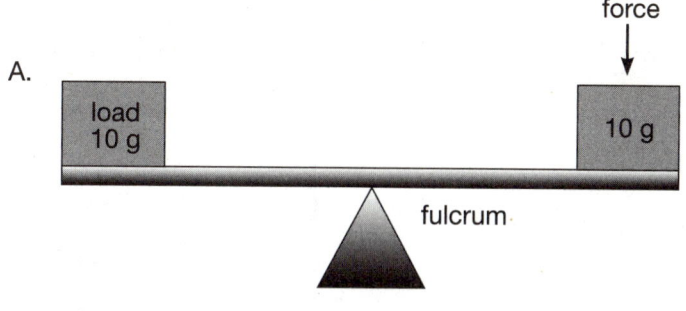

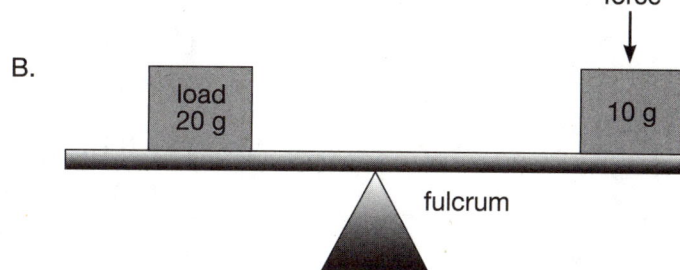

1. Compare the loads for each lever. How are the loads different?

   _____
   _____
   _____

2. Compare the distance between the fulcrum and the load of each lever. How are they different?

   _____
   _____

3. In lever B, the load is halfway between the fulcrum and the left edge of the lever. What is the difference between the force for lever A and the force for lever B? Explain.

   _____
   _____
   _____

 **Lesson 15** Levers, Pulleys, and Inclined Planes

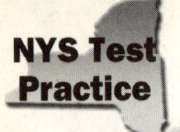 **DIRECTIONS** Choose the best answer for each question. Then circle the letter of the answer you have chosen.

1 Which is an example of an inclined plane?

A a lawnmower

B a wheelchair ramp

C an axe

D a bicycle wheel

2 What is the fulcrum in the picture?

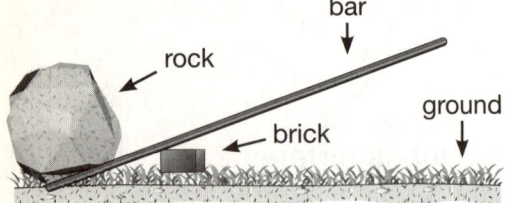

A the brick

B the rock

C the bar

D the ground

3 Which job could you do with a pulley?

A lift a box onto a truck

B pry a rock out of the ground

C pry a lid off a can

D split a log of wood

4  How does a fixed pulley make lifting a bucket full of cement easier?

A It increases the amount of force used.

B It changes the direction of the force.

C The force goes around a fulcrum.

D The force is increased by the bucket.

5 Which best describes the lever being used in the picture?

A The fulcrum is under the load.

B The force is above the load.

C The fulcrum is between the load and the force.

D The load is between the force and the fulcrum.

Lesson 16 — Gravity and Magnetism

Focus on the NYS Learning Standards

- **PS5.1c** The force of gravity pulls objects toward the center of Earth.
- **PS5.1e** Magnetism is a force that may attract or repel certain materials.
- **PS5.2a** The forces of gravity and magnetism can affect objects through gases, liquids, and solids.
- **PS5.2b** The force of magnetism on objects decreases as distance increases.

Learn how magnets attract and repel each other.

Gravity is a force that pulls objects to the center of Earth.

Something is **magnetic** if it can be pulled by a magnet from a distance.

Magnetic poles are the opposite ends of a magnet.

To **attract** is to pull toward something.

To **repel** is to push something away.

Guided Instruction

DIRECTIONS Read the following information.

Suppose you are playing kickball. You kick a ball into the air. It moves up because of the force of your kick, but then it falls down. The force that pulls it down is another kind of force called **gravity**.

Gravity does not need to touch an object to move it. It can pull through liquids or solids or gases. Gravity is the force that keeps the Moon and Earth close together. If not for gravity, the Moon would fly off into space.

Another force that can move objects without touching them is found in a magnet. You may know that some things will stick to a magnet and that some things will not stick. Paper clips and iron nails will stick to a magnet. Crayons or pencils will not stick. Objects only stick to magnets if they are **magnetic**. Magnets can also stick to other magnets.

Guided Questions

What happens to a kite that is affected by **gravity**?

What force keeps the Moon close to Earth?

Steel has iron in it. Is it **magnetic** or not **magnetic**?

Lesson 16 — Gravity and Magnetism

How do magnets stick? Magnets have two ends called **magnetic poles**. They are often just called poles. One end is called the north pole. The other end is called the south pole. On many magnets, the letter "N" marks the north pole and the letter "S" marks the south pole.

When the north pole of one magnet comes near the south pole of another magnet, they **attract**, or pull toward each other. Opposite poles attract each other.

When two north poles come near each other, the magnets push apart, or **repel**. The same thing happens when two south poles come together. In other words, poles that are alike repel each other.

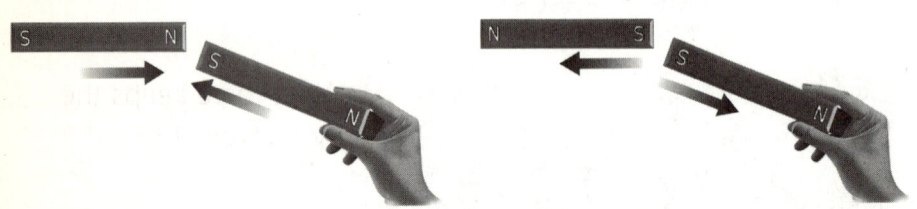

Magnets can attract objects even without touching them. They can attract objects through air, water, and even some solid things. You may have seen magnets holding a piece of paper on a refrigerator. The paper is not magnetic, but the refrigerator is. The magnet sticks to the refrigerator through the paper and holds it in place. In the same way, if you hold a magnet close to small pieces of iron, the iron pieces will move through the air toward the magnet.

When a magnet is far from something magnetic, it cannot pull on the object very well. The magnet's pull is weaker as it moves farther away. A magnet may stick to a refrigerator through a piece of paper, but it might not stick through ten pieces of paper.

Guided Questions

What are the two ends of a magnet called?

Circle the north poles on the magnets in the pictures to the left.

Which poles on a magnet **attract** each other?

Which **magnetic poles** on a magnet push apart?

How does distance affect a magnet's attraction?

DIRECTIONS For each question, write your answer in the spaces provided.

1. When you lift an object up, what force is pulling the object down at the same time?

   _____

2. What happens when unlike poles of two magnets come near each other?

   _____

3. What happens when the same poles of two magnets come near each other?

   _____

Lesson 16 — Gravity and Magnetism

Apply the NYS Learning Standards

DIRECTIONS Read the following paragraph and answer the questions.

Kim has four magnets. One of the magnets has its north and south poles labeled. She calls this magnet her Test Magnet. The other three magnets, shown here, do not have their poles labeled. Kim wants to figure out which end of each magnet is the north pole. She labels the ends of each magnet as shown here, so she can tell them apart.

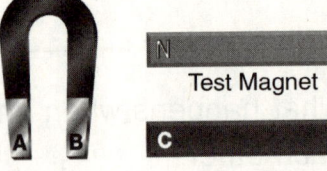

Magnet 1   Magnet 2   Magnet 3

Kim brings the north pole of the Test Magnet close to end A of Magnet 1. When she does this, Magnet 1 moves away from the Test Magnet.

1. Which end of Magnet 1 is the north pole? Explain your answer.

_____

_____

Kim brings the south pole of the Test Magnet close to end D of Magnet 2. When she does this, Magnet 2 moves away from the Test Magnet.

2. Which end of Magnet 2 is the south pole? Explain your answer.

_____

_____

Kim brings the north pole of the Test Magnet close to end F of Magnet 3. When she does this, Magnet 3 moves toward the Test Magnet.

3. Is end F of Magnet 3 a north pole or a south pole? Explain your answer.

_____

_____

4. If Kim brings end C of Magnet 2 near end E of Magnet 3, what will happen? Explain your answer.

_____

_____

Gravity and Magnetism **Lesson 16**

DIRECTIONS Choose the best answer for each question. Then circle the letter of the answer you have chosen.

1.

   The magnets shown above will most likely

   A move apart

   B break apart

   C become hot

   D move together

2. Which of these shows a pair of magnets that will repel each other?

   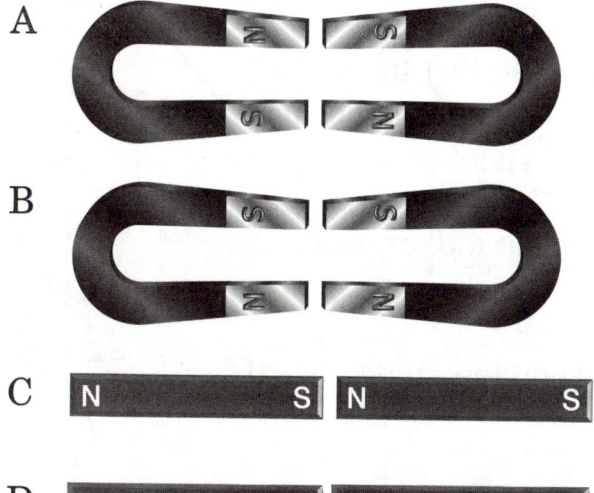

3. What causes rain and snow to fall?

   A friction

   B gravity

   C moving air

   D magnetic force

4. Which statement about magnets is true?

   A They only attract objects through air.

   B They must touch an object to attract it.

   C They must touch an object to repel it.

   D They can attract objects through water.

5. What happens when you move a magnet far away from an object that is attracted to it?

   A The magnet's ability to attract the object grows weaker.

   B The magnet's ability to repel the object grows stronger.

   C The object moves faster and faster.

   D The magnet's ability to attract the object grows stronger.

Performance Task

Make a Pulley

Focus on the NYS Learning Standards: T1.1a, T1.2a, T1.4a, b, T1.5a, b, c, ICT6.2, IPS7.1, IPS7.2

Task:

Build a simple pulley.

A pulley is a simple machine. It changes the direction of force needed to do work.

Materials:

- wire hanger
- 2 empty spools
- string
- 5-gram weight

Directions

1. Your teacher will give you a wire hanger. Carefully insert one spool through the open ends of the wire.

2. Squeeze the ends of the hanger together so that the loose ends connect. Bend the ends down so that the spool cannot come loose. Make sure you can turn the spool easily.

3. Hook the hanger onto a closet pole, curtain rod, or hook. The spool should be free to turn.

4. Tie one end of the string to a weight. Loop the string around the spool.

5. Pull the string to lift the weight.

6. In what direction is the force moving to lift the weight? In what direction does the weight move?

_____

7. Add a second spool to your pulley. Use a longer piece of string tied to the weight. Wrap the string as shown in this picture. Now you have a double pulley.

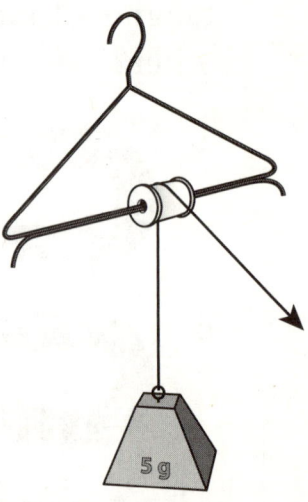

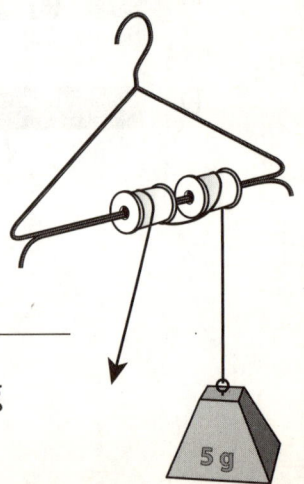

8. Now pull the string to lift the weight. What do you notice?

_____

_____

9. When would it be useful to have a pulley?

_____

10. What happens to the amount of force you need when you use a pulley?

_____

11. Here are some examples of pulleys in everyday life. Explain how each one works.

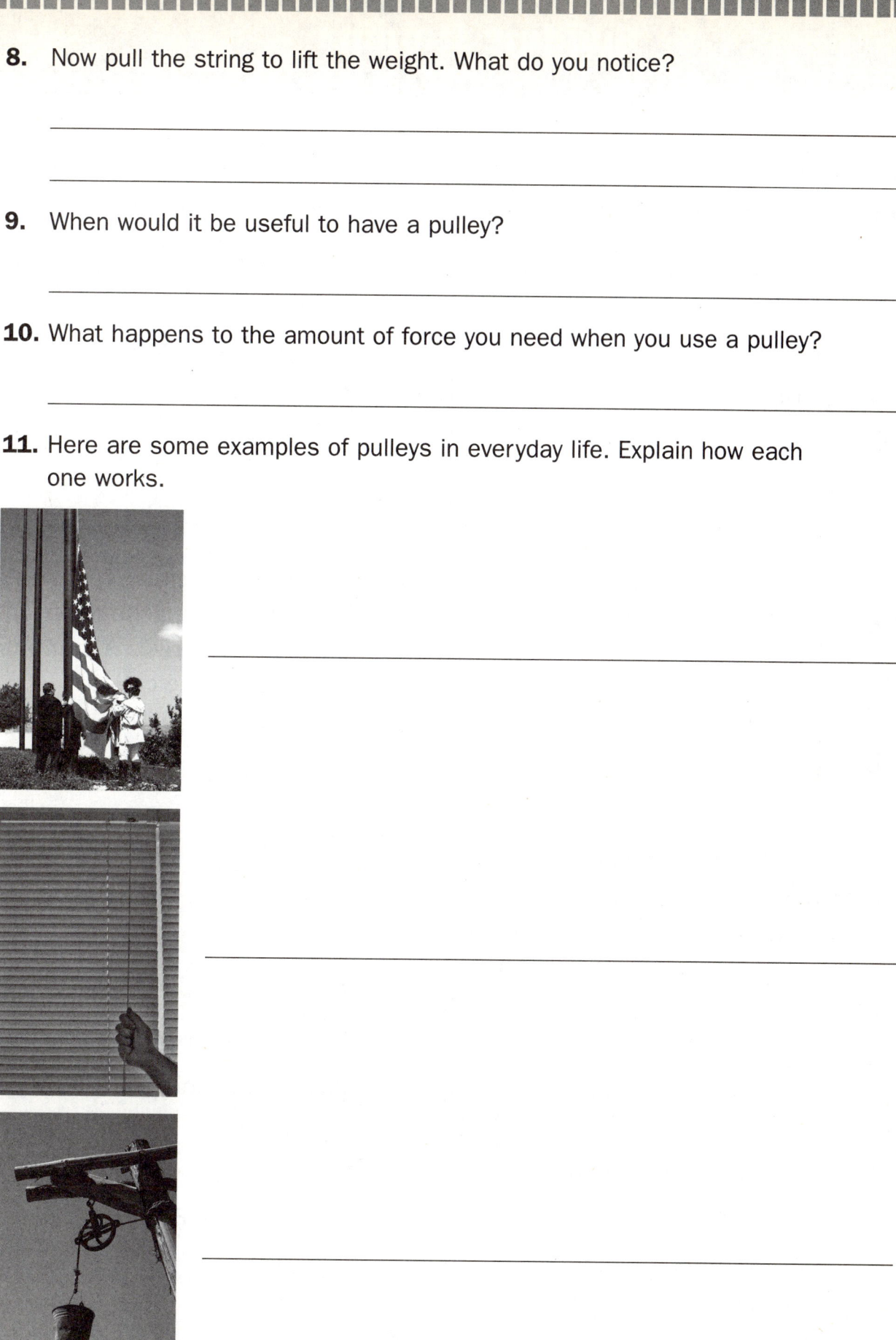

Building Stamina

Part I

Directions (1–14): Each question is followed by four choices. Decide which choice is the best answer. Circle the letter of the answer you have chosen.

1 Which causes a rocket to drop back to Earth?
   A speed
   B friction
   C gravity
   D wind

2 Motion is a change in
   A friction
   B position
   C direction
   D force

3 What makes the brakes on a bicycle work?
   A magnets
   B spokes
   C gravity
   D friction

4 Friction most helps a person
   A ski faster
   B spin around
   C slow down
   D jump up

5 Which of the following is not a force?
   A a push or a pull
   B gravity
   C a lake
   D friction

6 A delivery truck driver uses a ramp to move packages from the truck to the street. A ramp is a simple machine called
   A a pulley
   B an inclined plane
   C a lever
   D a fulcrum

7

   The rock is being lifted by
   A a pulley
   B a wedge
   C an inclined plane
   D a lever

8. Which of these jobs uses a lever?
   A raising a flag up a flagpole
   B removing a cap with a bottle opener
   C pushing a cart up a ramp
   D sweeping the floor

9.

   The machine lifting the bucket is called
   A a pulley
   B a fulcrum
   C an inclined plane
   D a lever

10. Which of these will most likely stick to a magnet?
    A
    B
    C
    D

11. You always place school work on the refrigerator with a magnet. Today, you use a magnet to hold 5 pieces of paper to your refrigerator. Then the magnet and the papers slide to the floor. Which most likely is the reason?
    A The refrigerator is not magnetic.
    B The papers are weak magnets.
    C The magnet is too strong to hold 5 pieces of paper.
    D The magnet is too weak to attract through 5 pieces of paper.

Building Stamina

12 The north pole of a magnet will stick to

A the north pole of another magnet.

B the south pole of another magnet.

C a piece of glass.

D a piece of plastic.

13 How would you describe the position of the flower?

A It is under the butterfly.

B It is on top of the butterfly.

C It is next to the butterfly.

D It is over the butterfly.

14 Which sentence is true about magnets?

A They can repel plastic and cotton.

B They only attract through solids.

C They can repel some gases.

D They can attract through liquids.

Part II

Directions (15–18): Record your answer on the lines provided below each question.

Suppose you want to lift a bucket. You have these materials.

  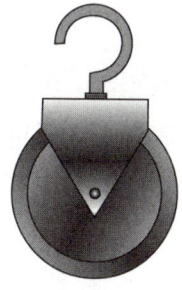

15 Tell how to put the materials together to lift the bucket.

_____

_____

16 Draw the simple machine you could make with the materials.

17 Explain how the simple machine helps you do work.

_____

18 How could you use the machine you drew to lift some heavy books?

_____

_____

Focus on the NYS Learning Standards

Chapter 5
Lesson 17 — Needs of Living Things

LE1.1a Animals need air, water, and food in order to live and thrive.
LE1.1b Plants require air, water, nutrients, and light in order to live and thrive.
LE1.1c Nonliving things do not live and thrive.
LE1.1d Nonliving things can be human-created or naturally occurring.
LE2.1a Some traits of living things have been inherited.

You can contrast living and nonliving things and identify the needs of living things.

A **living thing** grows, changes, and reproduces.

A **nonliving thing** does not grow, change, or reproduce.

To **reproduce** is to make more of the same kind.

A **trait** is a characteristic or feature that is passed down from one living thing to another.

Guided Instruction

DIRECTIONS Read the following information. Look at the pond below.

What do you see that is **living**? What do you see that is **nonliving**? How can you tell?

The frog can grow and change. It lays eggs and more frog's eggs hatch. It can **reproduce**. This is one way you know it is a living thing.

The plants you see also grow and change. They make seeds from which more plants grow. This is how they reproduce. They are living things.

The rocks do not grow. They do not reproduce. Rocks are nonliving things.

Guided Questions

How do the plants in the picture **reproduce**?

Needs of Living Things — Lesson 17

Living things have needs to stay alive. They must have air. The frog is breathing air. When it was a young tadpole, it got air from the water. The plants need air, and they need light, too. The light helps them grow. It helps them make food.

Animals and plants also need food and water. They need these things to grow. Do rocks need food and water? Do they need air and light? No. That is one more way you can tell they are nonliving things.

Rocks are found in nature, but they are not living things. Water can move, but it is not a living thing. Nonliving things can be found in nature. They can also be made by humans. Is a car a living thing? It moves, but it does not grow or reproduce. It needs gas to go, but it does not need water and food to grow. It is a nonliving thing.

You can group living things by looking at how they are alike. The frog's eggs will hatch. The tadpoles will grow into frogs. The new frogs will get many of their traits from their parents. This means that they will look like their parents. They will act like their parents in many ways. You can group frogs together as one kind of living thing.

The plant's seeds will grow. The new plants will look like the old plant. They will get their traits from that plant. Plants are very different from frogs. You would not group the two of them together. However, they are both living things.

Guided Questions

What three things do all living things need to stay alive?

Name some nonliving things in your classroom.

What traits might the new frogs get from their parents?

DIRECTIONS For each question, write your answer in the spaces provided.

1. Does a bird reproduce? Explain.

   _____

2. Besides food and water, what do plants need to stay alive?

   _____

3. How do you know that a pencil is nonliving?

   _____

Lesson 17 Needs of Living Things

DIRECTIONS Read the text below and answer the questions.

These things all look different. Some are animals, and some are plants. Some are nonliving things. How would you group them?

You can group the living things into animals and plants.

The animals all need air and food and water. They get these things in different ways. The rabbit eats plants. So does the snail. The stork eats small animals.

The plants all need air, light, food, and water. They get these things in different ways. They make their own food. Some plants need a lot of water. Some can get by with very little. Some need bright sunlight. Some can live with less light.

Each living thing needs a place to live that suits its needs. The right place will have enough food, water, light, and air for that living thing.

1. Circle the ones that need air, food, and water.

   _____

2. Make an X on the ones that need light as well as air, food, and water.

   _____

3. Name the nonliving things.

   _____

Needs of Living Things **Lesson 17**

NYS Test Practice

DIRECTIONS Choose the best answer for each question. Then circle the letter of the answer you have chosen.

1. What do animals need to live?
   A food, water, and light
   B water and food
   C air, food, and water
   D light and air

2. Why do plants need light?
   A to find food
   B to reproduce
   C to heat their leaves
   D to make food and grow

3. How do you know this is a nonliving thing?

   A It cannot move.
   B It has no eyes or ears.
   C It cannot think on its own.
   D It does not need air or food.

 4. What trait might a bear cub get from its parents?
   A fur
   B seeds
   C a big cave
   D berries and fish

 5. Which nonliving thing is found in nature?

   A

   B

   C

   D

Chapter 5 • Plants and Animals

Focus on the NYS Learning Standards

Lesson 18 Traits of Living Things

LE2.1a Some traits of living things have been inherited.
LE2.1b Some characteristics result from an individual's interactions with the environment and cannot be inherited by the next generation.
LE2.2a Plants and animals closely resemble their parents and other individuals in their species.
LE2.2b Plants and animals can transfer specific traits to their offspring when they reproduce.

You can understand that living things inherit traits that help them survive.

A **trait** is a body feature or behavior of a living thing.

A **species** is one kind of living thing.

An **adaptation** is a trait that helps a living thing meet its needs in its environment.

To **survive** is to remain alive.

To **reproduce** is to make more of the same kind of living thing.

Inherited traits are body features or ways of acting that are received from parents.

Learned behaviors are not inherited traits, but traits you must learn and practice.

Guided Instruction

DIRECTIONS   Read the following information.

Zebras live in the grasslands of Africa. They live in herds with other zebras. A trait of zebras is a striped coat. A **trait** is a characteristic. It describes body features or behaviors of living things.

The stripes on zebras help protect them from predators, such as lions. **Predators** are animals that hunt other animals for food. The stripes on a zebra make it hard for a lion to see one zebra in a herd. Some traits of a living thing may help protect it from predators. Other traits, such as the large teeth and sharp claws of a lion, can help an animal catch its food.

A **species** is one kind of living thing. For example, people belong to one species, and horses belong to another species. Living things of the same species can have similar traits that are slightly different.

Guided Questions

What is a **trait**?

What is a **species**?

Traits of Living Things   Lesson 18

On the grasslands of Africa, giraffes use their long necks and legs to reach and eat leaves from tall trees. Long ago, the ancestors of giraffes had necks and legs of different sizes. Some had shorter necks and legs. Others had longer necks and legs. They all ate leaves from trees. The giraffes with longer necks and legs could reach the leaves more easily.

A giraffe's long neck is an adaptation. An **adaptation** is a body feature or behavior that helps a living thing meet its needs in its environment. An adaptation is a way that a species adapts, or changes, to get food, water,

or shelter. Living things that adapt are more likely to **survive**, or remain alive. Giraffes with longer necks and legs survived. Living things that adapt are also more likely to **reproduce**. Animals reproduce when they give birth or lay eggs. Giraffes with shorter necks and legs died out, so they did not reproduce.

Another adaptation is the fur of the Arctic fox. This animal has a grey/brown coat in the summer. When winter comes, the fur of the Arctic fox changes to white. This helps them blend in with the environments of summer and winter. The animals they hunt are less likely to see them because of this.

Guided Questions

What **adaptation** helped the giraffe **survive**?

Lesson 18 Traits of Living Things

Tomato plants and pine trees are both types of plants. They have roots, stems, and leaves. They both use sunlight to make their food. These two kinds of plants look different from each other. They have different traits. The leaves of a pine tree are thin like a needle. The leaves of a tomato plant are wider with notches in them. A pine tree has a stem made of wood and is tall. The stem of a tomato plant is not as tall and is not made of wood. Can a tomato plant have needlelike leaves? Can a pine tree produce tomatoes? The answer to both questions is no.

Tomato Plant   Pine Tree

Guided Questions

This is because these plant traits are **inherited traits**. Inherited traits are body features or ways of acting that are passed from parents to their young. You have inherited your eye color or your height from one or both of your parents.

How an animal behaves is something else that can be inherited. How do baby turtles know to crawl to the ocean, and not toward land, after they hatch? This behavior is inherited from their parents. They did not have to learn this. A spider does not have to be taught to spin a web. Spiders inherit this behavior from their parents.

Where does a living thing get its **inherited traits**?

Traits of Living Things — Lesson 18

Not all the ways that animals act are inherited. Dogs can be taught to shake hands or roll over for food. A polar bear learns how to hunt seals by watching its mother. These behaviors are **learned behaviors**.

All living things need food. They need protection from predators. They also need traits to help them survive. Living things that have these useful traits will live and give birth to more living things. Those new living things will inherit the useful traits and will pass them on.

> **Guided Questions**
>
> Name a **trait** that dogs do not inherit from their parents.

DIRECTIONS For each question, write your answer in the spaces provided.

1. What are some traits that plants might inherit?

   _____

2. What is the difference between inherited traits and traits that are not inherited?

   _____

   _____

   _____

   _____

3. Suppose the branches of a tree are cut and then after that the tree looks different. Is this new tree shape an inherited trait? Why or why not?

   _____

   _____

   _____

Lesson 18 — Traits of Living Things

Apply the NYS Learning Standards

DIRECTIONS Read the paragraph, study the drawings, and answer the questions.

cocker spaniel   **husky**   **retriever**

The drawing shows three dogs: a cocker spaniel, a husky, and a retriever. All of these dogs are in the same species. There are many differences among dogs, such as how they look and act. Long ago, people bred dogs to be their companions and to help them do certain jobs. For example, people bred border collies to help them herd sheep.

1. What are some traits that are found in all dogs?

   _____
   _____

2. What are some of the traits that are different in dogs?

   _____
   _____
   _____

3. Explain which dog shown above is most likely to be a good hunting dog?

   _____
   _____

4. Name another type of dog and describe ways that it differs from the three dogs pictured at the top of the page.

   _____
   _____

Traits of Living Things | **Lesson 18**

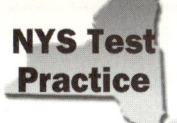

 DIRECTIONS Choose the best answer for each question. Then circle the letter of the answer you have chosen.

1. The inherited trait that would be most helpful to a winning racehorse is
   A long fur
   B long legs
   C short tail
   D good eyesight

2. From where does a plant get its inherited traits?
   A the Sun
   B its parents
   C the plants that grow around it
   D the kind of soil it is growing in

3. Which of these is not an inherited trait in a pea plant?
   A the color of its leaves
   B the shape of its leaves
   C the color of its flowers
   D the amount of water it gets

4. Which of these is a trait that a dog inherits from its parents?
   A fetching a stick
   B begging for food at the table
   C body size and shape
   D barking when it wants to go out

5. The drawing shows the leaf of a parent tree.

Which of the following shows an offspring leaf from the parent tree?

A

B

C

D

Chapter 5 • Plants and Animals

Focus on the NYS Learning Standards

Lesson 19 Life Cycle of Plants

LE4.1a Plants have life cycles.
LE4.1b Each kind of plant goes through its own stages of growth and development that may include seed, young plant, and mature plant.
LE4.1c The length of time from beginning of development to death of the plant is called its life span.
LE4.1d Life cycles of some plants include changes from seed to mature plant.
LE4.2a, b, LE5.1a, b

You can see changes over time in the growth of plants.
To **sprout** is to begin to grow.
A **life cycle** is made up of the changes an organism goes through.
The **life span** of a plant is the length of time from the beginning of its development to its death.
Fertilization is the first stage of a plant's growth.
A **seed** is a plant part that has a young plant and food stored inside.
A **seedling** is a young plant that has sprouted.

DIRECTIONS Read the following information.

Have you ever seen a plant change as time goes by? Have you ever planted a seed and watched it **sprout** and grow? All plants have a **life cycle**. A life cycle includes several stages of growth and change throughout the life of an organism. In fact, a plant will keep growing as long as it lives. For a bristlecone pine, that could mean more than 4,000 years! That length of time between the beginning of the plant's development and its death is called its **life span**.

It may be hard to imagine plants needing food, but they do. Food is important for all living things. Food gives living things the energy to grow. Food also helps living things repair themselves.

Guided Questions

What are the changes that a plant goes through during its life called?

When does a plant stop growing?

Why do living things need food?

Life Cycle of Plants — Lesson 19

Plants make their own food. They use sunlight to change material from the air into food they can use. They release some of the material they do not need into the air as oxygen. We use this oxygen when we breathe.

Different plants have different traits that help them survive. Plants in dry places have long roots that reach down to get water. Plants in dark forests have long trunks that let their leaves reach up for sunlight.

The life cycle of a plant begins with **fertilization**. Fertilization creates a tiny young plant, called an embryo. This embryo is wrapped in a protective coat with stored food called a **seed**.

Seeds will only sprout if the conditions are right. If a seed has oxygen, water, and the right temperature, it will hold water and swell. Tiny roots appear first, and then a tiny stem appears. The young plant uses the stored food to grow. The new roots grow down into the soil to take in water. The new stem grows upward toward the light. Soon the stem has a few leaves growing from it. The new young plant is a **seedling**.

A sprouting seed — seed coat, stored food, embryo, new root

When new leaves begin to grow, they use sunlight to make food for the plant. If there is enough sunlight, the new plant will continue to grow and change. It will get more leaves and roots. The stem will grow larger and taller. Some plants will produce flowers. These flowers will produce new seeds. The seeds will grow into new plants and begin the life cycle again.

Guided Questions

What is a **seedling**?

Why do plants need light?

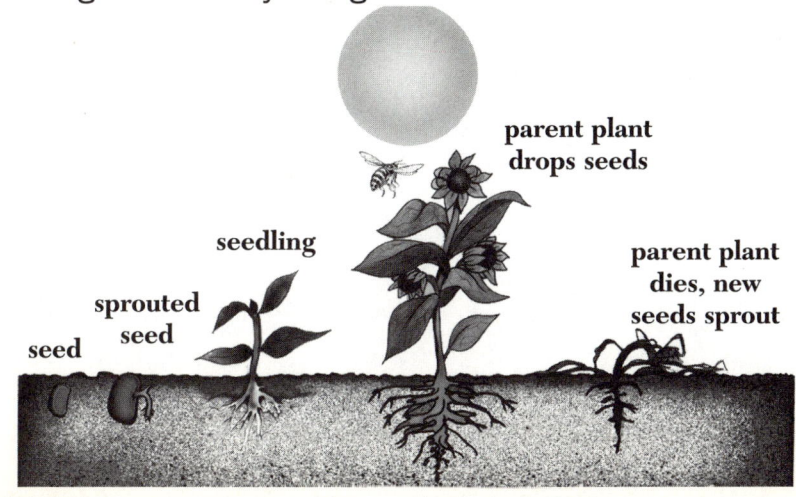

seed • sprouted seed • seedling • parent plant drops seeds • parent plant dies, new seeds sprout

Lesson 19 Life Cycle of Plants

The parent plant may continue to produce flowers and seeds for many years. The life span of that plant ends when it dies. But the life cycle of that kind of plant continues because it produced seeds.

Guided Questions

How do you know a plant's **life cycle** is over?

DIRECTIONS For each question, write your answer in the spaces provided.

1. What three things are needed for a seed to sprout?
   _____
   _____

2. What does the embryo inside a seed use for food?
   _____
   _____

3. Describe the stages in the life cycle of an oak tree.
   _____
   _____
   _____
   _____
   _____

Life Cycle of Plants **Lesson 19**

DIRECTIONS Read the paragraphs and answer the questions.

Two students set up an experiment. They want to test how sunlight affects the growth of bean seedlings. They place dampened paper towels in two bags with zip tops. They place ten bean seeds on top of the paper towels in each bag. They close the bags halfway. One of the bags they wrap completely in aluminum foil. They leave the other uncovered. Then they place the two bags on a sunny windowsill. Every day for ten days, they check the bean seeds and observe their growth. Whenever the paper begins to dry out, they moisten it again.

The students observe that all seeds sprouted. But once they sprouted, only the seedlings with light have bright green leaves. The stems are short and sturdy. The seedlings without light have small, yellowish leaves. The stems are long and spindly. The students conclude that bean seedlings need sunlight to grow properly.

1. What do you predict would happen to the seeds in the dark after another week or two?

_____

_____

2. Why did the students place the seeds on dampened paper towels?

_____

_____

3. What do you think would have happened if the students had placed the bags in the refrigerator?

_____

_____

4. The students plant the bean seedlings in soil. They place them in a warm, sunlit area and water them regularly. What will happen?

_____

_____

_____

Lesson 19 Life Cycle of Plants

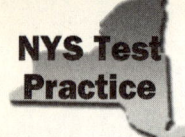

DIRECTIONS Choose the best answer for each question. Then circle the letter of the answer you have chosen.

1. Which drawing shows the correct order of the life cycle of the sunflower plant?

   A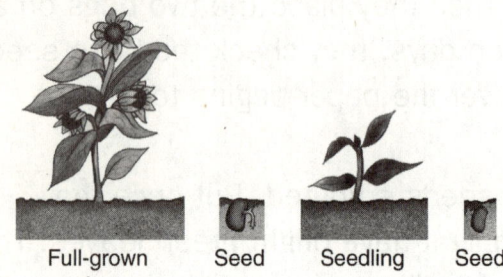
   Full-grown plant — Seed — Seedling — Seed

   B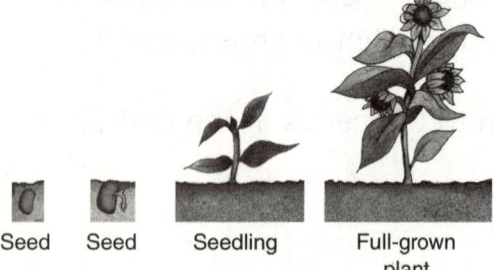
   Seed — Seed — Seedling — Full-grown plant

   C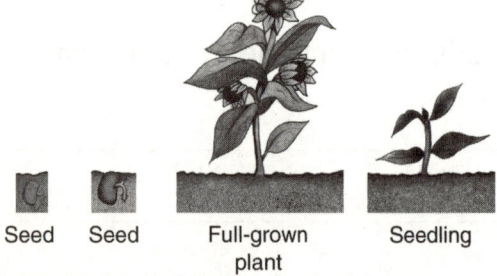
   Seed — Seed — Full-grown plant — Seedling

   D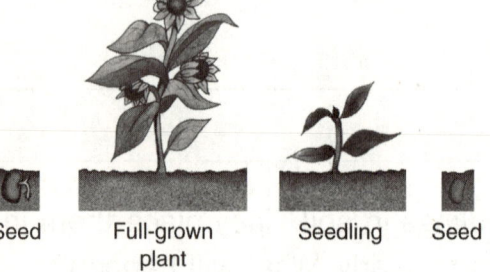
   Seed — Full-grown plant — Seedling — Seed

2. A plant stops growing
   A when it dies
   B after one year
   C after ten years
   D after fertilization

3. The life cycle of a plant begins with
   A leaves
   B a seedling
   C a sprouting seed
   D fertilization

4. What do plants do that humans do not do?
   A use food
   B use oxygen
   C reproduce
   D make oxygen

5. A tiny young plant inside a seed is called the
   A embryo
   B seedling
   C sprout
   D flower

Lesson 20 Life Cycle of Animals

Focus on the NYS Learning Standards

LE4.1e Each generation of animals goes through changes in form from young to adult. This completed sequence of changes in form is called a life cycle. Some insects change from egg to larva to pupa to adult.
LE4.1f Each kind of animal goes through its own stages of growth and development during its life span.
LE4.1g The length of time from an animal's birth to its death is called its life span. Life spans of different animals vary.
LE4.2a, b, LE5.1a, b

You can observe changes in animals during their life cycles.
A **larva** is the early, immature stage in the life cycle of some animals.
The **pupa** stage of metamorphosis is when a larva changes into an adult.
Metamorphosis is a change in body form during the growth and development of some animals.

Guided Instruction

DIRECTIONS Read the following information.

What do a butterfly and an elephant have in common? They go through changes during their life cycle. The changes a butterfly goes through are different from those of an elephant.

The butterfly begins its life cycle as an egg. When the egg hatches, a caterpillar comes out. The caterpillar is a **larva**. This is an early stage in the life cycle of the butterfly. The larva eats and grows. Then it forms a chrysalis, often called a cocoon, around itself. The chrysalis looks like a tiny sack.

The caterpillar becomes a **pupa** inside the chrysalis. The pupa slowly becomes a butterfly. When the butterfly is completely formed, the chrysalis breaks open. The adult butterfly comes out.

The butterfly goes through **metamorphosis**. Its body changes form as it grows from an egg to an adult. Many insects go through metamorphosis. The adult butterfly mates and lays eggs to begin the life cycle again. When the adult butterfly dies, its life span is over.

egg

larva

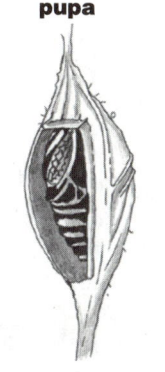

pupa

adult

Guided Questions

What is the first stage in the life cycle of a butterfly?

What happens to a butterfly's body as it goes through **metamorphosis**?

Lesson 20: Life Cycle of Animals

An elephant does not go through metamorphosis. A baby elephant looks a lot like an adult elephant. Elephants grow as they get older. However, their basic shape does not change. Elephants of all ages have a head with big ears, a long trunk, a body, and four legs.

Butterflies and elephants have very different life spans. A butterfly may live just one month after its metamorphosis. An elephant may live 70 years. Larger animals often have a longer life span.

Humans go through the same kind of life cycle that elephants do. Many other animals have this kind of life cycle. This includes other mammals, birds, reptiles, and fish. These animals look almost the same throughout their entire lives.

All animals share some things in common. They all breathe air. They all need to take in food to grow and to repair cuts and bruises. They all give off waste materials. All animals go through a life cycle, too. They grow through different stages. They can have their own young when they become adults.

Guided Questions

How is an elephant's life cycle different from a butterfly's?

How is an elephant's life span different from a butterfly's?

Life Cycle of Animals **Lesson 20**

DIRECTIONS For each question, write your answer in the spaces provided.

1. What happens to a caterpillar after it forms a chrysalis?

   _____

   _____

2. During which stage in a butterfly's life cycle are eggs laid?

   _____

3. What kind of a life cycle does an elephant have?

   _____

   _____

4. What are life functions that all living things perform?

   _____

   _____

5. Choose your favorite animal and research its life cycle stages. Using the space below, draw three stages of the animal's life cycle.

Lesson 20 Life Cycle of Animals

DIRECTIONS Read the information, study the drawings, and answer the questions.

Frogs are amphibians. Amphibians also go through metamorphosis. But their changes are different from those of a butterfly, as shown below.

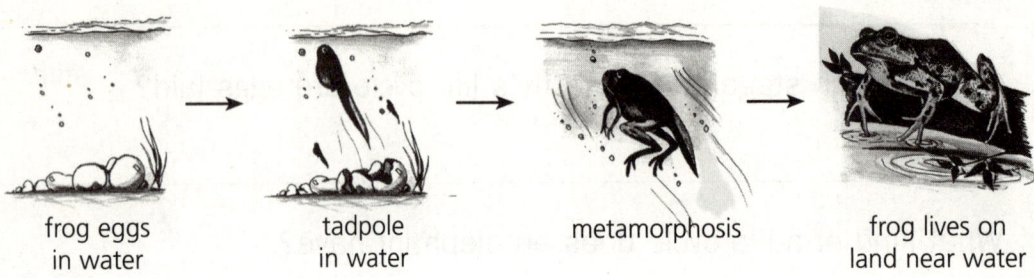

frog eggs in water → tadpole in water → metamorphosis → frog lives on land near water

1. Where do tadpoles live?

   _____

2. Where do frogs live?

   _____

3. Animals that breathe air need lungs. Animals that can get air from water use gills. Do you think a tadpole changes into a frog on the inside, too? Explain your answer.

   _____
   _____
   _____

4. How are the life cycles of a frog and a butterfly different? How are they alike?

   _____
   _____
   _____

5. What two major changes can be seen as a tadpole changes into a frog?

   _____

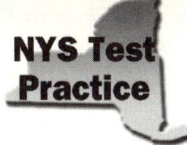

Life Cycle of Animals **Lesson 20**

NYS Test Practice

DIRECTIONS Choose the best answer for each question. Then circle the letter of the answer you have chosen.

1 Which stage shows the larva of the silkworm moth?

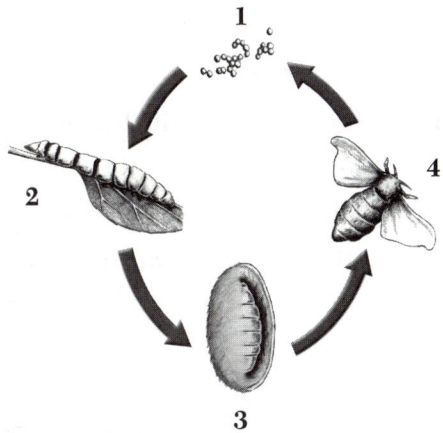

- A 1
- B 2
- C 3
- D 4

2 Which two animals have life cycles that are the most alike?
- A butterfly and crocodile
- B butterfly and chicken
- C frog and human
- D cat and elephant

 3 Which is true of all animals?
- A They go through metamorphosis.
- B They use lungs to breathe air.
- C They need food to grow.
- D They have a short life span.

 4 Which would you expect to have the longest life span?
- A hummingbird
- B earthworm
- C dog
- D human

5 Tadpoles have gills because they
- A must get air from water
- B live on land near ponds
- C have a short life span
- D change their shape

Peoples Education Copying is illegal. Chapter 5 • Plants and Animals 113

Focus on the NYS Learning Standards

Lesson 21 Adaptations of Plants

LE1.1b Plants require air, water, nutrients, and light in order to live and thrive.
LE3.1b Each plant has different structures that serve different functions in growth, survival, and reproduction.
LE3.1c In order to survive in their environment, plants and animals must be adapted to that environment.
LE5.2a Plants respond to changes in their environment. For example, the leaves of some green plants change position as the direction of light changes; the parts of some plants undergo seasonal changes that enable the plant to grow; seeds germinate; and leaves form and grow.
LE1.2a, LE4.2a, b, LE5.2g, LE6.1d, f

You can identify and compare the traits that help plants survive.

An **environment** includes all the living and nonliving things that surround an organism.

An **adaptation** is a behavior or part of a living thing that helps it survive.

Pollen is a powdery substance produced in flowers that helps plants reproduce.

Guided Instruction

DIRECTIONS Read the following information.

The **environment** of an organism includes all the living and nonliving things that surround it. **Adaptations** are special behaviors or parts of living things that help them survive. Like all living things, plants grow, use food and air, reproduce, and give off waste in order to survive. Many kinds of plants have adaptations, such as roots, leaves, stems, flowers, and seeds. These adaptations help them survive. Different plants in the same environment often have similar adaptations. They respond to the same conditions.

Plants in a tropical rain forest respond similarly, due to the conditions there. In a tropical rain forest, temperatures are always warm. Rain falls almost every day. Plants and trees in a rain forest have many adaptations for survival. The many kinds of plants in the tropical rain forest form layers. The tallest trees grow above all the other trees and plants. These trees have straight, smooth trunks and few branches. The leaves are all at the top, above all the other plants. This helps them get plenty of sunlight. They use the sunlight to make their own food.

Guided Questions

Why do plants living in the same **environment** often have similar **adaptations**?

Adaptations of Plants — Lesson 21

The trees in a tropical rain forest have very shallow roots. This is because most of the nutrients are in the top layer of the soil.

Cactus plants are common in the desert where it is hot and dry. The roots of the saguaro cactus are just below the surface. They stretch far in every direction. This allows the plant to soak up a great deal of water even if there is only a tiny amount of rain. The cactus also has small spiny leaves and a waxy stem. Both of these adaptations help prevent water loss.

Oak trees grow in temperate forests. These forests get plenty of rain, but they have cold winters. The trees must survive freezing temperatures. In the autumn, the oak leaves drop off. If they stayed on the trees in winter, the water in them would freeze. This would damage these trees and make it easier for diseases to attack. Also, without leaves, oak trees do not need as much water during winter.

So plants adapt to their environments, but sometimes environments change. When that happens, plants must adapt or die. If water becomes hard to find, plants must grow deeper roots. If sunlight is hard to reach, they must grow taller, or they must turn to face the Sun.

Some plants have flowers. The flower of a plant helps the plant reproduce. Each flower produces fruit that contains seeds. From the seeds, new plants grow.

Some birds and insects help plants reproduce. They move **pollen** from flower to flower. Pollen is a powdery substance produced in flowers that helps plants reproduce.

A plant will always grow toward the Sun.

Guided Questions

What are three **adaptations** of the tallest trees in the tropical rain forest?

Why does a saguaro cactus have roots so close to the surface?

What is the main problem for plants that live in temperate forests?

How does a flower help a plant survive?

Lesson 21 Adaptations of Plants

Flowers adapt to attract those birds or insects. Some flowers are shaped like tubes. Those flowers attract birds with long beaks or butterflies with long tongues. Some flowers have flat pads that allow bees to walk right in. Some flowers smell like rotten meat. That attracts flies.

Seeds adapt, too. A plant can survive if it reproduces. It must spread its seeds far and wide. Some seeds are adapted to fly in the air. Some are adapted to stick to passing animals. Those seeds can travel far. New plants will grow where the seeds fall to Earth.

All plant adaptations are designed to help plants survive. The adaptations help plants to grow. They help them to use sunlight to make food. They help them to reproduce. As long as plants get the air, water, space, and sunlight they need to grow, make food, and reproduce, they will survive. If the environment changes, they must adapt or die.

Guided Questions

What are some **adaptations** that help a plant reproduce?

DIRECTIONS For each question, write your answer in the spaces provided.

1. What is a reason that oak trees in temperate forests need adaptations to survive the winters?

   _____
   _____
   _____

2. How are the adaptations of a tropical rain forest tree and a desert cactus alike? Why?

   _____
   _____
   _____

Adaptations of Plants — Lesson 21

Apply the NYS Learning Standards

DIRECTIONS Read the text below and answer the questions.

Stems function to support a plant, and they carry water and food. Stems produce new plant cells. Stems vary greatly. They provide many different adaptations for plants.

- A potato plant has an underground stem (a tuber) that stores food for the plant. It also can produce new plants without seeds from each of the "eyes" on the potato.
- Strawberry plants have long stems called "runners" that grow sideways instead of upward. At the end of each one of these runners, a new strawberry plant can grow.
- A celery stalk is a stem that stores food for the plant.
- A tree trunk is also a stem. It carries water and food up and down the tree. Its hard outer bark provides protection from insects and parasites. It also keeps water inside the tree.
- A water lily stem has large air spaces in it. These air spaces keep the plant light and elastic, so it can float in the water. The floating stems connect to leaves that float on the surface, where they can get sunlight to make food.

1. What are two types of stems that store food for a plant?

   _____

2. How is a tree trunk different from other kinds of stems?

   _____
   _____

3. Why would a water lily not survive if its stems did not have air spaces?

   _____
   _____
   _____

4. What are three functions that are basic to stems?

   _____
   _____

Lesson 21 — Adaptations of Plants

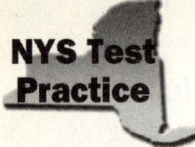

DIRECTIONS Choose the best answer for each question. Then circle the letter of the answer you have chosen.

1. Which statement about plants is not true?
   - A They make their own food.
   - B They reproduce.
   - C They can live without water.
   - D They need air to survive.

2. A change in the environment can lead some plants to
   - A grow in the dark
   - B float on water
   - C move away
   - D adapt and survive

3. Without sunlight, plants cannot
   - A make food
   - B take in water
   - C attract insects
   - D spread seeds

 4. Some plants have special adaptations of their stems that allow them to do all of the following except
   - A store food
   - B store water
   - C take in water from deep underground
   - D produce new plants without seeds

 5. Which plant is most likely to have large air spaces in its stems to help it get air?
   - A a desert plant
   - B a water plant
   - C a tropical rain forest plant
   - D a temperate forest plant

 6. Which adaptation is most likely to help a plant stay alive through a cold winter?
   - A shallow roots
   - B an underground stem
   - C dropping its leaves in fall
   - D supports extending out from its base

Focus on the NYS Learning Standards

Lesson 22 Adaptations of Animals

LE3.1a Each animal has different structures that serve different functions in growth, survival, and reproduction.
LE3.1c In order to survive in their environment, plants and animals must be adapted to that environment.
LE5.2e Particular animal characteristics are influenced by environmental conditions. These behaviors may include: fat storage in winter, coat thickness in winter, camouflage, shedding of fur.
LE5.2f Some animal behaviors are influenced by environmental conditions. These behaviors may include: nest building, hibernating, hunting, migrating, and communicating.
LE1.1a, LE5.1b, LE5.2b, c, d, g, LE4.2a, b

You can identify and compare the traits that help animals survive.

A **trait** is a feature of a plant or animal.

An **adaptation** is a trait that helps a plant or animal survive.

A **predator** is an animal that hunts and eats other animals.

Prey are animals that predators hunt and eat.

The **gills** of a fish allow it to breathe underwater.

Some animals **migrate**, or adapt to change by traveling.

Guided Instruction

DIRECTIONS Read the following information.

Giraffes have long necks. Polar bears have heavy fur coats. These special features, or **traits**, help the organisms stay alive and survive.

All animals have basic needs that must be met for them to survive. Animals need food, water, air, and a place to live. They also need protection.

An **adaptation** is a trait that helps an organism meet its needs. For example, giraffes, polar bears, and fish have many adaptations. A giraffe's long neck helps it reach leaves on high branches that most other animals cannot reach. This means that giraffes do not have to compete for food with shorter animals. The giraffe has a spotted coat, a long fringed tail, and a long tongue. A giraffe's tail acts as a flyswatter to keep flies and other pests away. The giraffe's long tongue helps strip leaves off trees. The spots of different sizes and colors on its coat help hide the giraffe, especially when it is young, from lions and other **predators**. Predators are animals that hunt other animals for food.

Guided Questions

What is a special feature of an organism called?

What are the basic needs of all organisms?

What do **adaptations** help organisms do?

Lesson 22   Adaptations of Animals

Giraffes are adapted to live on the plains of Africa. Giraffes could not live in the icy arctic north, and they could not live under water. But other animals are adapted to live in these areas.

Polar bears have thick layers of blubber, or fat. This helps them stay warm. The polar bear's heavy fur coat also helps it stay warm in the icy arctic north. Polar bears are predators. They hunt mostly seals for their food. They have excellent senses of smell and sight to help them find their prey. The animals that a predator hunts are called **prey**. Polar bears are white, which helps them blend in with the snow. This is important for the cubs, so that they can hide from predators.

Most fish have a pointed front, a wider middle, and a tapered tail. This shape makes fish streamlined. It helps fish move more easily through water. A fish has **gills** that allow it to take in oxygen from the water. Fish have fins instead of legs. The fins move them and guide them through the water.

Many fish are dark on top and light on the bottom. This makes them hard to see from either above or below and protects them from predators.

Animals have adapted to their environments. If the environment does not supply the things an animal needs, the animal may not be healthy enough to grow and live.

What happens when the environment changes? Sometimes there are unusual changes, such as forest fires and floods. Animals then must move to find a new place to live. Then there are changes that happen every season. It gets very hot or very cold. It gets very dry or very rainy. Food may be hard to find. Some animals adapt to that kind of change by traveling. Birds may **migrate** by flying north or south. Elk may migrate up or down a mountain.

Guided Questions

What are three **adaptations** of polar bears?

How does a fish's shape help it?

What might make animals **migrate**?

Other animals have other ways of adapting. Bears hibernate for part of the winter when food becomes scarce. They sleep in a safe place, often a cave. Turtles, snakes, and chipmunks hibernate, too.

What happens to you when your environment changes? If it gets very hot, you may perspire. If it gets very cold, you may shiver. If it is very sunny, you may squint or blink. If you are high up on a mountain, you may breathe faster. All adaptations have the same purpose. They are all designed to help living things survive.

> **Guided Questions**
>
> What are some changes that happen to humans with a change in environment?

DIRECTIONS For each question, write your answer in the spaces provided.

1. What will happen to an organism if one of its basic needs is not met?

   _____

2. How do a polar bear's adaptations help it stay alive?

   _____

   _____

   _____

3. What are two adaptations of a giraffe that help it meet the basic need of finding food?

   _____

   _____

Lesson 22 Adaptations of Animals

Apply the NYS Learning Standards

DIRECTIONS Read the paragraph, study the drawings, and answer the questions.

All birds have feathers, but different birds eat different kinds of foods. Each kind of bird has a beak that is adapted for getting food. Woodpeckers pound holes in trees with their long beaks and pull out insects. Some ducks scoop up tiny plants and animals from the water. Finches find and crack open tiny seeds. Eagles tear apart prey, such as mice and rabbits. Each of these kinds of birds has a beak that is an adaptation. Their beaks help them eat specific kinds of foods and help them to survive.

Woodpecker **Duck** **Finch** **Eagle**

1. Why do birds have beaks that are different shapes?

   _____

   _____

2. What is the shape of a finch's beak? How does this help the finch?

   _____

   _____

3. What is the shape of an eagle's beak? How does this help the eagle?

   _____

   _____

4. A hummingbird is a tiny bird that sips sugary nectar from inside tube-shaped flowers. What shape do you think its beak is? Why?

   _____

Adaptations of Animals — Lesson 22

DIRECTIONS Choose the best answer for each question. Then circle the letter of the answer you have chosen.

 1 Which foot below most likely belongs to a bird that has adapted to paddling through the water?

A

B

C

D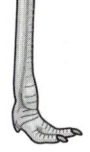

 2 Harp seals give birth to their young on snowy ice fields in the Arctic. The baby seals have white coats. The mothers go off and leave their young alone. How does its white coat help a baby seal stay alive?

A It makes it easier for the mother to find it.

B It makes it hard for predators to see it.

C It helps it break through chunks of ice.

D It helps it get air from the surface.

3 Lack of food might lead some animals to

A grow

B shed fur

C build nests

D migrate

4 Some animals might shiver in response to

A cold

B dryness

C lack of food

D darkness

 5 In a grassland, buffalo eat grass. What might happen if there was a fire in the grassland?

A The grass would never grow back.

B There would be less food for the buffalo, so they would have to hibernate.

C The buffalo would have to adapt and become predators.

D There would be less food for the buffalo, and they would have to migrate.

Focus on the NYS Learning Standards

Lesson 23 — Response and Behavior

LE3.2a Individuals within a species may compete with each other for food, mates, space, water, and shelter in their environment.
LE3.2b All individuals have variations, and because of these variations individuals of a species may have an advantage in surviving and reproducing.
LE5.2a, b, c, e, f, LE6.1

Living things in a habitat compete for resources.
Living things **compete** when they struggle against each other for things from their habitat.
Hibernate is what some animals do when they sleep through the winter.
Camouflage is a body shape or color that helps an animal blend into its surroundings.

Guided Instruction

DIRECTIONS Read the following information.

Living things need resources from their habitats in order to live. Each plant needs resources such as space, light, air, and water. Each animal needs resources such as oxygen, water, food, shelter, and space. Sometimes, a habitat does not have enough resources for all the animals or plants that live in it. The living things in such a habitat have to **compete** for the resources. Living things compete when they struggle against other living things to get the resources they need.

Not all living things compete against each other. For example, bears and trees have different needs. They are not likely to compete against each other.

Animals that need the same resources are more likely to compete against each other. Often, animals of the same population compete. For example, black bears in a forest may have to compete against other black bears in the same forest for food. They may compete for water, space, shelter, and mates, too. The bigger, stronger bears will get more resources than the smaller, weaker bears.

Animals use their senses to react to danger. If they hear a sound, they can run away. If they see another animal come too close to them, they can scare it away or run. If they smell the smoke of a forest fire, they run, too.

Guided Questions

Name some resources that living things need.

What do living things do when they **compete**?

124 Science • Level C Copying is illegal. Measuring Up® to the New York State Learning Standards

Animals also react to the change of seasons. In the spring, birds build nests. In the winter, many birds fly

south where it is warm and where they can find food. Other animals dig a deep hole under the ground and hibernate. **Hibernate** means to sleep through the winter. Animals do this because there is no food in the winter. Chipmunks and groundhogs hibernate.

Changes in the environment cause some animals' behaviors to change. Their traits can change, too. Remember how the Arctic fox's coat turns white in the winter? That was so it would blend in with the snow. Blending in with the environment is called **camouflage**. Animals have other ways to survive the cold winters, too. Some animals, such as polar bears, store extra fat for the winter. Many animals grow thicker fur in the winter. Cats and dogs grow a thicker fur in the winter. Then, in the warmer spring, they shed some of their fur.

Plants respond to changes in the environment, too. You know that many trees lose their leaves in the autumn. In the winter, the trees do not have leaves and they do not grow, but they are alive. They are just resting and waiting for warmer weather. Did you know that the growth of trees also slows down when there is a drought in the summer? Other plants, such as grasses, will turn brown when it is dry. But as soon as there is rain, the grass will turn green.

Guided Questions

Why do chipmunks **hibernate**?

Describe how an animal would look if it were **camouflaged**.

How do some trees respond to the coming of winter?

Lesson 23 Response and Behavior

DIRECTIONS For each question, write your answer in the spaces provided.

1. What resources would two black bears compete for if they lived in the same area?

   _____

   _____

2. Why would a bigger, stronger bear be able to survive better than a smaller, weaker bear?

   _____

   _____

3. How would an animal know if there is another animal or danger in its environment?

   _____

   _____

4. What are two things that animals can do to survive the winter.

   _____

   _____

5. How do some trees survive the cold winter?

   _____

   _____

Response and Behavior **Lesson 23**

Apply the NYS Learning Standards

DIRECTIONS Read the paragraphs and answer the questions.

Many animals live in deserts. These animals include bobcats, coyotes, rattlesnakes, Gila monsters, iguanas, wrens, woodpeckers, and jackrabbits. However, there are very few trees in the desert. Birds have few places to build nests. It is hard for them to find food. At night, they need protection from larger animals.

Desert habitats receive little rainfall—less than 25 centimeters per year. That means there is very little water in the desert for all the animals. Animals that live in the desert compete for water. Where do they get the water they need? They get most of their water by eating desert plants that store water. They also get water by eating other animals, as shown in the picture.

1. Desert animals must find water. How do they get water?

   _____
   _____

2. Why is there so little water in a desert?

   _____
   _____

3. There are few trees in the desert. What does this mean for birds living there?

   _____
   _____

Lesson 23 Response and Behavior

NYS Test Practice

DIRECTIONS Choose the best answer for each question. Then circle the letter of the answer you have chosen.

1. Several trees are growing close together in a forest. What does this most likely mean?
   A   None of the trees will survive.
   B   No other trees will grow in this forest.
   C   The trees will have a lot of space in which to grow.
   D   The trees will have to compete for space, water, and sunlight.

2. What would most likely happen if a bird population in a forest got very large?
   A   There would be more competition for resources.
   B   There would be more resources for everyone to use.
   C   There would be less competition for resources.
   D   There would be more habitats in the forest.

3. Which is an animal behavior that is caused by colder weather?
   A   thicker fur
   B   hibernation
   C   camouflage
   D   extra fat

4. Which animal would best survive in a place where digging for grubs is the main means of finding food?
   A   the one with the longest claws
   B   the one with the biggest mouth
   C   the one with the quickest feet
   D   the one with the thickest fur

5. If a deer could see and smell that there is no food left in one field, the deer would most likely respond by
   A   hunting other deer
   B   living on water
   C   moving to another field
   D   sleeping all winter

6. Which change in the environment might make animals respond by moving away to find resources?
   A   rainfall
   B   sunlight
   C   nightfall
   D   forest fire

Performance Task

Water Moves Through Stems

Focus on the NYS Learning Standards: LE1.1b, LE3.1b, S3.2a, S3.3a, S3.4a, T1.3a, T1.4b, T1.5b, ICT6.6, IPS7.2

All plants need water. Plants move water from their roots through their stems. The inside of the stem has parts that cause the water to move up into the leaves and flowers of the plant.

Task:

You will use simple materials to observe how a plant moves water through its stem.

Materials:

- 2 white carnations
- 3 plastic cups
- 2 different colors of food coloring
- marker
- water
- colored pencils
- magnifying glass

Procedure:

1. Use the marker to label the cups A, B, and C.

2. Fill each cup exactly halfway with water. Use the marker to draw a line at the water level of each cup.

3. Add 15 drops of one food coloring to Cup A. Add 15 drops of the other food coloring to Cup B.

4. Place 1 carnation in Cup A and 1 in Cup B.

5. What do you think will happen to the flowers in Cups A and B?

_____

_____

6. Do you think the water level in Cups A and B will change in 4 hours? Will the water level in Cup C change? Explain your answers.

_____
_____
_____

7. Complete the table below. In the correct spaces, draw your observations for the start of the experiment, after 4 hours, and for the next day. Use colored pencils to draw the cups, water, and flowers. Don't forget to draw a line showing the water level in each cup.

	Cup A	**Cup B**	**Cup C**
Start			
4 hours			
Next day			

8. What happened to the water levels of Cups A, B, and C?

_____

9. Did your observations match what you thought would happen? Write down what you observed at the end of the performance task.

_____
_____
_____
_____

10. Explain why you think this happened.

_____

_____

_____

_____

_____

_____

11. What do you think would happen to the color of a white flower if you placed one in Cup C? Why?

_____

_____

_____

12. Cut each flower stem in half so that you have a top half and a bottom half. Observe the cross-section with a magnifying glass. Use colored pencils to draw what you see.

13. When you have finished, follow your teacher's instructions for cleaning up your work area and putting everything where it belongs. Dry any objects that are wet and wipe up any spills.

Building Stamina

Part I

Directions (1–15): Each question is followed by four choices. Decide which choice is the best answer. Circle the letter of the answer you have chosen.

1 Which of the following is false?
   A Seeds can be many sizes and shapes.
   B An apple tree grows from an apple seed.
   C A seed grows to look like the plant it came from.
   D Two different kinds of seeds will grow into plants that look alike.

2 What is it called when living things struggle against each other to meet their need for food?
   A erosion
   B ecosystem
   C competition
   D environment

3 A young plant is a
   A seed coat
   B seedling
   C root
   D stem

4 Which two animals have life cycles that are the most alike?
   A frog and chicken
   B human and elephant
   C butterfly and dog
   D frog and human

5 Which animal pictured below is a larva at one stage in its life cycle?

   A    C

   B    D

6 Which of the following traits is not inherited?
   A blue eyes and curly hair
   B large hands and feet
   C short height
   D large muscles from lifting weights

7

Which of these birds has the strongest beak best suited for cracking and eating hard nuts and seeds?

A   finch

B   sea bird

C   sparrow-like bird

D   pigeon

8   A plant's roots are bent from growing around a rock to reach water. Which statement best describes the plant's roots?

A   The shape of the roots is inherited.

B   The shape of the roots is the result of a learned behavior.

C   The shape of the roots is the result of adapting to the environment.

D   The roots were not affected by the rock.

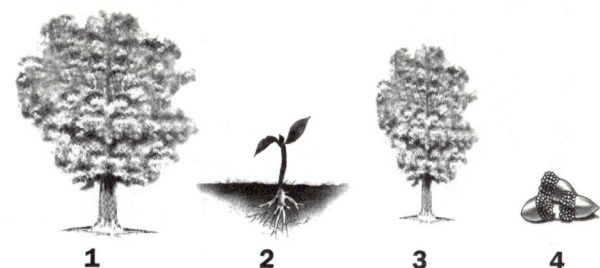

9   The pictures show different steps in the life cycle of an oak tree. Which is the correct order for the steps in the life cycle?

A   1, 2, 3, 4

B   4, 2, 3, 1

C   4, 3, 1, 2

D   3, 4, 2, 1

10   Which of the following is a learned behavior?

A   A fox preys on a rabbit.

B   A frog jumps when touched.

C   A bird builds a nest out of grass.

D   A dog returns to where it found food.

11   How can you describe food, water, and protection for living things?

A   traits

B   behaviors

C   basic needs

D   adaptations

12 In which environment is an animal most likely to need adaptations to drink seawater?

A Drawing A

B Drawing B

C Drawing C

D Drawing D

13 Which is not true of all living things?

A They get air from water.

B Their life span ends in death.

C They get rid of waste.

D They may reproduce.

14 Which of the following best describes why polar bears and giraffes have different traits?

A Polar bears rely mostly on inherited traits.

B Giraffes rely mostly on learned traits.

C They live in different environments.

D They have different basic needs.

15 Animals need energy to grow and heal. This energy comes from

A air

B food

C the Sun

D Earth

Part II

Directions (16–20): Record your answer on the lines below each question.

Many plants and flowers have tubes in their stems to move food and water. A student filled Cups A and B with water. In Cup B, the student also placed 15 drops of blue food coloring. Then the student put one white carnation in each cup.

16 What would you expect to happen? Finish the table with the results that the student might see.

	CUP A	CUP B
Start	white carnation in 1 cup of clear water	white carnation in 1 cup of blue water
Next Day		

17 Why would the water level change?

_____

18 Predict how the flowers might look the next day.

_____

_____

_____

19 Some flower shops sell green carnations. They did not spray them green. How did they make them?

_____

20 Suppose you bought a green carnation. Your friend asked, "Will it lose its color if you put it in clear water?" How could you test to find out?

_____

Focus on the NYS Learning Standards

Chapter 6
Lesson 24 Living Things Need Energy

LE5.1a All living things grow, take in nutrients, breathe, reproduce, and eliminate waste.
LE5.1b An organism's external physical features can enable it to carry out life functions in its particular environment.
LE6.1a Green plants are producers because they provide the basic food supply for themselves and animals.
LE6.1b All animals depend on plants. Some animals eat other animals.
LE6.1d Decomposers are living things that play a vital role in recycling nutrients.
LE6.2a Plants manufacture food by utilizing air, water, and energy from the Sun.

You will find out how different living things get the energy they need.

Producers make their own food from sunlight, water, and air. Plants are producers.

Consumers eat other living things for food. Animals are consumers.

Carnivores are consumers or animals that eat other animals.

Herbivores are consumers or animals that eat only plants.

Omnivores are consumers or animals that eat both other animals and plants.

Decomposers get energy by breaking down decayed plants or animals.

Guided Instruction

DIRECTIONS   Read the following information.

Living things get energy from food. Not all living things get food from the same places. Plants get energy from the Sun. Then they make their own food by using sunlight, water, and air. Plants are called **producers**. The food that they produce is stored in the leaves, roots, seeds, and fruit of the plant.

Animals cannot make their own food like plants do. Animals are called **consumers**. They get energy by eating plants or other animals. All of the living things in the picture below are consumers except the leaf. The leaf is part of a producer. There are different types of consumers. We classify or group them by the kinds of food they eat.

Guided Questions

What is a **producer**?

What is a **consumer**?

What is a **carnivore**?

Which organisms in the picture are probably **carnivores**?

owl

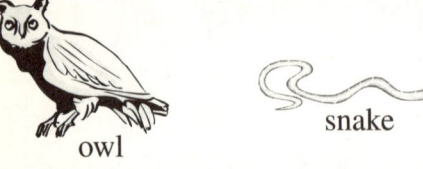
snake

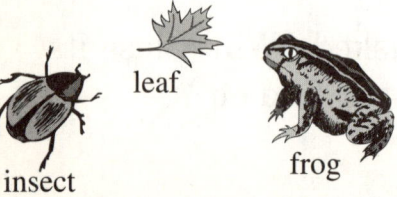
leaf   frog

insect

Consumers that eat only animals are called **carnivores**. Lions, wolves, and sharks are all carnivores with sharp teeth. Some kinds of insects are carnivores, too. These insects eat other insects for food. But most insects eat plants.

Living Things Need Energy — Lesson 24

Consumers that eat only plants are called **herbivores**. The food that is stored in the different parts of plants becomes food for herbivores. This passes energy along to the animals that eat the plants. Cows, rabbits, sheep, goats, panda bears, and deer are all herbivores.

Pandas eat bamboo plants.

Consumers that eat both plants and animals are called omnivores. Most people are **omnivores**. The meat that people eat comes from animals such as fish, chickens, pigs, or cows. The fruits and vegetables that people eat come from plants.

A living thing that gets energy from nonliving things is a **decomposer**. A decomposer breaks down decaying plants and animals into small pieces. These pieces get mixed with soil and help plants to grow. Mushrooms and earthworms are decomposers.

Guided Questions

What is an **herbivore**?

What is an **omnivore**?

DIRECTIONS For each question, write your answer in the spaces provided.

1. How does a producer get its food?

   _____

2. What are three types of consumers?

   _____

3. What type of animal is most likely to have sharp teeth: a carnivore, herbivore, or an omnivore? Explain.

   _____

4. Do carnivores eat consumers or producers or both? Explain.

   _____

   _____

Lesson 24 Living Things Need Energy

Apply the NYS Learning Standards

DIRECTIONS Use the drawings to answer the questions.

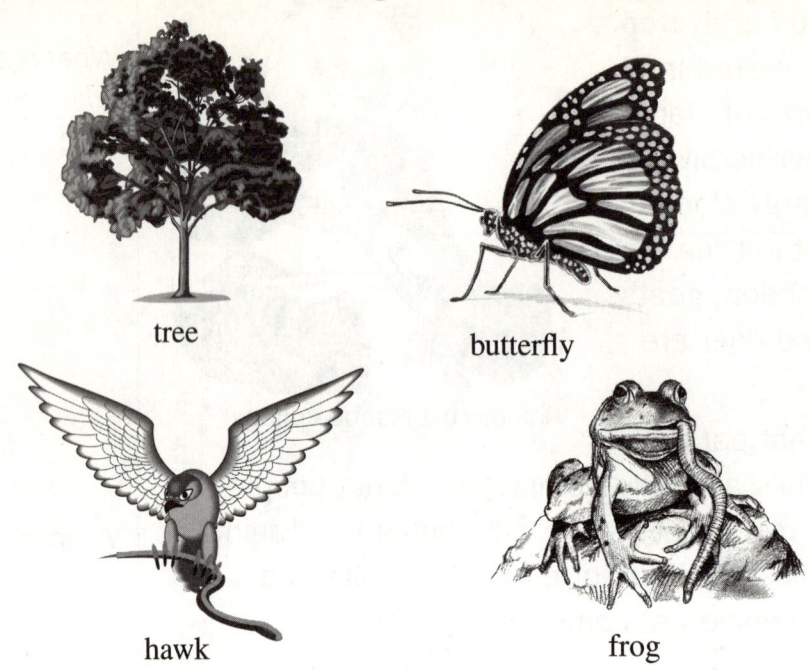

tree

butterfly

hawk

frog

1. Classify each living thing above as a consumer or a producer.

   _____
   _____
   _____
   _____

2. Which two living things could be carnivores? Why?

   _____
   _____

3. Earthworms eat dead plants and animals. How can we classify the earthworm that the frog is eating?

   _____
   _____

4. Most butterflies drink nectar from flowers for food. How can we classify butterflies?

   _____
   _____

Living Things Need Energy **Lesson 24**

NYS Test Practice

DIRECTIONS Choose the best answer for each question. Then circle the letter of the answer you have chosen.

1. Which of the following is not a consumer?

   A   robin

   B   rosebush

   C   snake

   D   goldfish

2. Why are decomposers important to the environment?

   A   They break down decayed plants and animals and help plants grow.

   B   They make their own food from sunlight.

   C   Carnivores eat only decomposers for food.

   D   Herbivores eat only decomposers for food.

3. Where do producers get their energy from?

   A   animals

   B   plants

   C   the Sun

   D   Earth

4.

   The organism in the picture is a(n)

   A   herbivore

   B   decomposer

   C   carnivore

   D   producer

5.

   Why might a scientist think that this animal skull came from a carnivore?

   A   Only carnivores have skulls.

   B   It has a large jaw.

   C   It has a flat head.

   D   It has sharp teeth.

Focus on the NYS Learning Standards

Lesson 25 Food Chains

LE6.1b All animals depend on plants. Some animals eat other animals.
LE6.1c Animals that eat plants for food may in turn become food for other animals. This sequence is called a food chain.
LE6.2b The Sun's energy is transferred on Earth from plants to animals through the food chain.

You will learn how food passes from plants to animals to other animals in a food chain.

A **food chain** is the path of food from one living thing to other living things.

A **predator** is an animal that hunts and eats other animals.

Prey is an animal that is hunted by other animals for food.

Guided Instruction

DIRECTIONS Read the following information and answer the questions.

You know that plants use energy from the Sun to make their own food. Plants are eaten by some animals. These animals may then be eaten by other animals.

Food travels from the producers to the consumers to the decomposers. This is called a **food chain**.

Animals that hunt other animals for food are **predators**. Most predators are carnivores such as cats, wolves, and sharks. Other predators are omnivores. The animals that predators hunt are **prey**.

Sometimes an animal can be both a predator and prey. Look at the forest food chain shown below. The arrows show the path of food. The tree is a producer. The beetle eats the leaves of the tree. The frog eats the beetle. The snake eats the frog. Finally, the owl eats the snake.

In this food chain, the frog and the snake are both predators and prey. The beetle is a prey. The owl is a predator. It is at the top of the food chain.

Guided Questions

What is a **food chain**?

Which animals hunt other animals for food **predators** or **prey**?

What living thing is at the beginning of a forest **food chain**?

In a **food chain**, in which direction do the arrows go?

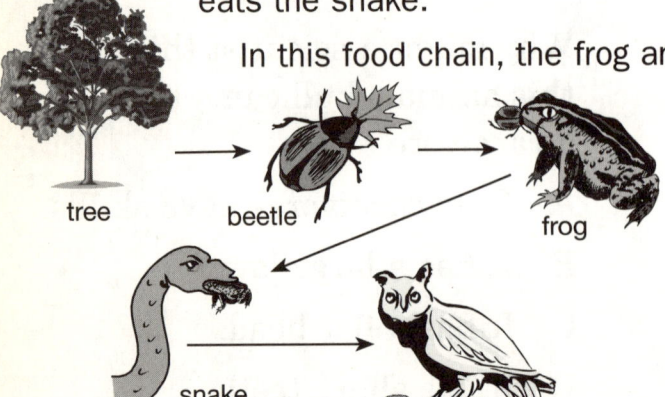

forest food chain

Food Chains — Lesson 25

Lakes, rivers, and oceans all have different plants and animals in their food chains. Algae and plants such as seaweed are the main producers for food chains in water. Look at the ocean food chain shown here. The fish feed off of the seaweed. Predators, such as the barracuda and the shark, hunt and eat the smaller fish.

The barracuda is both a predator and prey. The shark is the carnivore at the top of this food chain.

Changing one part of a food chain also changes the other parts. If the algae and plants in the ocean food chain stopped growing, there would be no food for the fish. If the fish died, the barracudas and sharks would not have food.

Guided Questions

Which **predator** in this **food chain** is also **prey**?

DIRECTIONS For each question, write your answer in the space provided.

1. How can an animal be both a predator and prey?

   _____

   _____

2. Why are plants at the beginning of food chains?

   _____

   _____

3. What might happen if one animal was taken out of a food chain?

   _____

   _____

4. Where are people in a food chain?

   _____

   _____

Lesson 25 Food Chains

Apply the NYS Learning Standards

DIRECTIONS Use the drawing to answer the questions.

grass   mouse   snake   owl

1. Which of the animals in this food chain are predators?
   _____

2. Which of the animals in this food chain are prey?
   _____

3. Which living thing is at the top of the food chain?
   _____

4. Owls can eat mice as well as snakes. What would happen if the snake in the food chain were to disappear?
   _____
   _____

5. How does food get from the plant to the owl in this food chain?
   _____
   _____

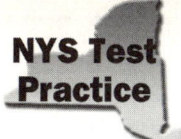

NYS Test Practice

DIRECTIONS Choose the best answer for each question. Then circle the letter of the answer you have chosen.

1. The sea lion in this picture is
   A   prey
   B   a predator
   C   an herbivore
   D   a producer

2. What type of animal is most often found at the top of a food chain?
   A   producer
   B   decomposer
   C   herbivore
   D   carnivore

 Which living thing completes this food chain?

   plant → ? → fish → bird → alligator

   A   shark
   B   seaweed
   C   hawk
   D   insect

4. Which living thing in this food chain is both a predator and prey?
   A   grass
   B   insect
   C   snake
   D   owl

5. Where do all food chains get their energy from?
   A   from predators
   B   from the Sun
   C   from prey
   D   from algae

Focus on the NYS Learning Standards

Lesson 26 Humans and Their Environments

LE5.2d Some animals, including humans, move from place to place to meet their needs.

LE5.3a Humans need a variety of healthy foods, exercise, and rest in order to grow and maintain good health.

LE5.3b Good health habits include hand washing and personal cleanliness; avoiding harmful substances; eating a balanced diet; engaging in regular exercise.

LE7.1a Humans depend on their natural and constructed environments.

LE7.1b Over time humans have changed their environment by cultivating crops and raising animals, creating shelter, using energy, manufacturing goods, developing means of transportation, changing populations, and carrying out other activities.

LE7.1c Humans, as individuals or communities, change environments in ways that can be either helpful or harmful for themselves and other organisms.

You can explore how humans adapt to and change their environments.

A **shelter** is a place that protects a living thing.

An **environment** is everything in and around the place where a plant or animal lives.

A **community** is a group of living things that work and live together.

To **cultivate** is to grow plants for a purpose.

To **adapt** is to change to fit a certain environment.

Guided Instruction

DIRECTIONS Read the following information.

What do humans need to live? They need air and sunlight. They need food and water. They need clothing and **shelter**.

The earth is very large. It has many different **environments**. Some environments are very cold. Others are very hot. Some are very wet. Others are very dry. Humans live in most parts of the earth. How do they do it?

Long ago, humans lived along rivers and streams. They lived where they could easily get water and food. Over time, they gathered in groups. They built communities.

Guided Questions

Why is **shelter** important for humans?

Where did human **communities** first appear?

144 Science • Level C Copying is illegal. Measuring Up® to the New York State Learning Standards

Humans and Their Environments — Lesson 26

At first, humans ate what they could hunt or gather. Later, they began to **cultivate** crops. They grew food to eat. They used animals to do work. They built boats to travel. Sometimes, these changes changed the environment. Humans cut down trees to grow crops. They let their animals eat the plants in the fields. They dug channels for their boats.

Like all animals, humans moved to find food. They moved to find water or to escape bad weather. They traveled far to find what they needed. They built new communities.

As they moved to new places, humans had to **adapt**. They had to change what they ate. They might not be able to find that kind of food or grow those crops in the new place. They had to change what they wore. Their old clothing might be too warm or not warm enough. They had to change the kind of shelter they built. They might not find the same materials for building. They might need a different kind of home.

Today, you can find humans living all over the earth, like in the places shown below. Humans depend on their environment to give them what they need, like food and water. They build shelters to protect them from heat or cold or rain.

Guided Questions

Why did humans **cultivate** crops?

What are some reasons humans moved?

Why did humans need to **adapt** to their new environments?

DIRECTIONS For each question, write your answer in the spaces provided.

1. Why did humans first live along rivers and streams?

2. What are three ways in which early humans changed their environment?

3. Choose one of the environments above. Tell what you might need to do to adapt to that environment.

Lesson 26 Humans and Their Environments

DIRECTIONS Read the following information.

Some environments are hard on humans. The Sahara Desert in Africa is very hot and dry. You can see the heat waves in the picture below. Every year, the desert grows a little more. Humans must move to find water.

The humans here raise sheep and goats. They drink their milk and eat their meat. They use their skins to build shelters. When the animals eat all the plants, the humans must move on.

This is a hard way to live. Because of that, many people who once lived this way have moved to cities. In cities, they can find food and water more easily. They do not have to build a new shelter every few months. They must stop raising sheep and goats, though. They must adapt to find new jobs.

In many parts of the world, cities are growing very fast. Humans are moving from hot, cold, wet, or dry environments where it is hard to live. Larger communities have greater needs. As the cities grow, they are changing the environment around them.

1. What makes the Sahara Desert a hard place to live?

   _____

2. How do humans use the sheep and goats they raise?

   _____

3. Why are so many humans moving from the countryside to the city?

   _____

4. What does the author mean when she says "larger communities have greater needs"? Explain.

   _____

Humans and Their Environments **Lesson 26**

NYS Test Practice

DIRECTIONS Choose the best answer for each question. Then circle the letter of the answer you have chosen.

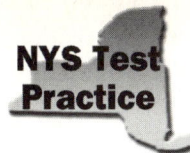

1 Raising animals changed humans' environment because

   A animals drank the water the humans needed

   B animals ate the plants in the environment

   C animals needed to live near rivers and streams

   D animals cut down trees in the forests

2 Which was not a major reason for early humans to move?

   A to find food

   B to escape bad weather

   C to find water

   D to get a new job

3 How did cultivation change humans' diets?

   A It added more meat to their diets.

   B It took away meat from their diets.

   C It gave them a new source of vegetables.

   D It meant they no longer ate fish.

4 Why do humans need shelter?

   B It protects them.

   B It feeds them.

   C It comforts them.

   D It does work for them.

5 How did the earliest humans get food?

   A by buying and selling

   B by raising sheep and goats

   C by cultivating crops

   D by hunting and gathering

6 The changes needed for cultivation might be harmful to

   A horses and cows

   B forest animals

   C sea creatures

   D desert plants

Performance Task

Investigate Air Quality

Focus on the NYS Learning Standards: M2.1a,b, M3.1a, S2.3a, b, S3.2a, S3.4a, b, T1.1b, c, T1.2c, ICT6.1, ICT6.2, ICT6.4, IPS7.2

Task:

You will work with a partner to investigate the air quality in one place at your school. Then, you will share results with your class.

Materials:

- 10 × 15 cm index card
- 12 cm strip of clear packaging tape
- hand lens or magnifying glass
- clothes pin or small weights to keep the card in place
- metric ruler
- pencil
- scissors
- sheet of white paper

There are many things that can get into the air and make it dirty. Some of these things are invisible gases that mix with the air. Others are tiny specks that you can see if you look closely. Cars and factories can cause air pollution by releasing harmful things into the air. However, things such as dust can also make the air dirty. For example, think of what the air looks like after a car drives down a dirt road. In this activity, you will explore tiny specks that get into the air.

Directions:

1. As a class, brainstorm different things that can get into the air. Think of times when you have seen tiny specks in the air. Write the class responses in the spaces below.

_____

_____

_____

2. Work with a partner. You and your partner will use an index card and packaging tape to make a tool for collecting tiny specks in the air. Your teacher will make this tool along with you. Be sure to ask for help if you are unsure what to do.

3. Fold the 10 × 15 centimeter an index card in half, like a greeting card.

4. Keep the index card folded in half. Use your pencil and ruler to draw a line that is 4 centimeters long along the center of the folded edge. Then, draw lines that are 5 centimeters long straight down from the ends of the first line. Connect these lines to make a rectangle.

5. Keep the index card folded. Use scissors to cut out the rectangle you drew. Start cutting at the folded edge and cut through both layers of the index card. This will make a window in the index card when you unfold it.

6. Unfold the index card. Write your name in one corner.

7. Lay the 12-centimeter strip of packaging tape on your desk with the sticky side UP. Carefully place the paper on top of the tape so that the tape shows through the window. **Be very careful NOT to touch the tape or stick the tape to anything.** Touch ONLY the card.

8. You will use the index card to collect tiny specks in the air near your school. Work with your teacher to choose a place to test. Write the place you chose in the space below. Some examples: near the playground, near the street, in your classroom, in the art room, in the bathroom, near the front door, or near the baseball field.

_____

9. Your teacher will create a classroom chart with three columns. You and your partner will write your names in the first column. Write the place you chose for your card in the second column.

10. Make a hypothesis about which place you think will have the most pollution in the air. Write your hypothesis in the space below.

_____

_____

11. Everyone in the class made the same tool with the same size opening and the same amount of tape. Why is it important that all the index cards are the same?

_____

_____

12. Your teacher's index card will go into a clean container with a lid. What is the purpose of placing the card there?

13. When your teacher tells you, bring your card to the place you chose. Use a clothes pin or a small weight, such as a rock, to keep your card in place. Leave your card there for 24 hours.

14. Why are all the cards left for the same amount of time?

    _____

    _____

15. After 24 hours, bring your index card back to the classroom.

16. Make sure that your card has the sticky side UP. Place the non-sticky side on a piece of white paper. Use a hand lens to count the number of specks on the tape that is stuck to the card. Record the number of specks in the space below.

    _____

17. In the third column of the class chart, you and your partner will write the number of specks that you counted next to your name.

18. Which card had the most specks? Why do you think this was the case?

    _____

    _____

19. How many specks were on the card in the closed container? Why do you think this was the case?

    _____

    _____

20. Did the results of the investigation support your hypothesis?

    _____

21. Imagine that you were going to do this performance task again. What would you do differently?

    _____

    _____

    _____

Building Stamina

Directions (1–14): Each question is followed by four choices. Decide which choice is the best answer. Circle the letter of the answer you have chosen.

1. Why are both the tree and the bush called producers?

   A They remove dead plants from the environment.

   B They make their own food.

   C They depend on animals for food.

   D They depend on other plants for food.

2. Which of these best describes what an omnivore might eat?

   A grass and insects

   B algae and seaweed

   C frogs and insects

   D bird eggs and mice

3. The snake in this picture eats the jackrabbit. What is true about the snake?

   A It is a predator.

   B It is an herbivore.

   C It is a decomposer.

   D It is both a producer and a consumer.

4. Which one would you most likely find at the end of a food chain?

   A decomposers

   B herbivores

   C producers

   D prey

5 Which of these is an herbivore?

A

B

C

D

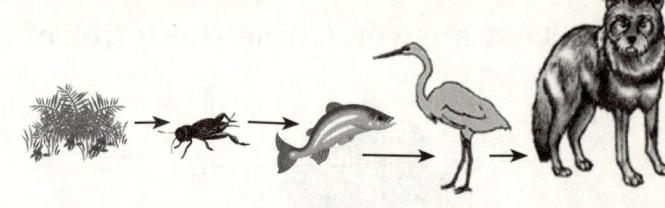

grass   grasshopper   fish   egret   coyote

6 Except for the grass, all of the organisms in the food chain shown above are

   A predators
   B producers
   C prey
   D consumers

7 Where do producers get energy?

   A from prey
   B from Earth
   C from consumers
   D from the Sun

8 Early humans often moved from place to place to

   A find food
   B learn new skills
   C meet new people
   D escape predators

9. How do carnivores get energy from the Sun?
   A by eating only plants
   B by living outdoors
   C through the food chain
   D by acting as producers

10. How does cultivation usually change the environment?
    A It covers wetlands with dry soil.
    B It moves from mountains to valleys.
    C It changes the paths of rivers.
    D It replaces trees with cropland.

11. In addition to food and water, what do humans need?
    A prey
    B shelter
    C weather
    D trees

12. Which one is a decomposer?
    A a mushroom breaking down a dead log
    B a large fish eating a smaller fish
    C a peanut plant from which many products are made
    D a rock lying at the bottom of a small stream

13. Which does not name a way that human change can be harmful to other organisms?
    A Building a road can cut through a deer's habitat.
    B A large factory can pollute the air with smoke.
    C Raising grains can feed livestock.
    D Cultivation can remove a small forest habitat.

14. Hunting and gathering is one way a human
    A acts as a predator
    B eats a balanced diet
    C acts as a producer
    D makes his or her own energy

Directions (15–17): Record your answer on the lines provided below each question.

The people of Townsville cleaned up Townsville Lake. Students wanted to know whether the cleanup worked. They tested the water before and after the cleanup. Students scooped up water and let it settle for 24 hours. Then they measured to find the amount of solid pollution that settled to the bottom. This chart shows one student's findings.

	Solids settled in water
Before the Cleanup	Nearly $\frac{1}{2}$ inch
After the Cleanup	Less than $\frac{1}{16}$ inch

15 Which of these tools would work best for the test? Explain your answer.

  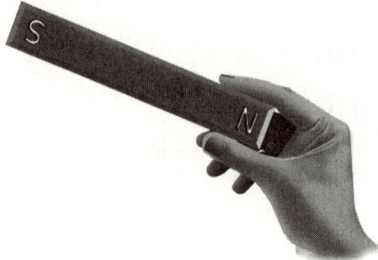

_____

_____

16. Most of the students had findings like the one student on the previous page. Based on this, would you say that the cleanup had an effect? What was it? Explain your answer.

_____

_____

17. Some students at the north end of the lake had these findings.

	Solids settled in water
Before the Cleanup	Over $\frac{1}{2}$ inch
After the Cleanup	Nearly $\frac{1}{2}$ inch

What question would that lead you to ask?

_____

_____

_____

Focus on the NYS Learning Standards

Chapter 7
Lesson 27 Weather

PS2.1a Weather is the condition of the outside air at a particular moment.
PS2.1b Weather can be described and measured by temperature, wind speed and direction, form and amount of precipitation, and general sky conditions (cloudy, sunny, partly cloudy).
PS2.1c Water is recycled by natural processes on Earth.

You can explain how forecasters predict the weather.
A **thermometer** is a tool that measures temperature.
A **wind vane** is a tool that shows the direction of the wind.
An **anemometer** is a tool that measures wind speed.
A **rain gauge** is a tool that collects and measures amounts of rainfall.
Fog is a cloud that touches the Earth's surface.

Guided Instruction

DIRECTIONS Read the following information.

What kind of weather do you have today? Is it sunny? Is it raining? What kinds of clouds are in the sky? Weather is the condition of the air outside. It changes from day to day. If you observed weather changes every day, after a while you would notice patterns in the weather. Then you would be able to predict weather yourself.

You can measure elements of the weather. You can measure temperature by using a **thermometer**. You could find the wind direction by using a **wind vane**. You can use an **anemometer** to measure wind speed. You can use a **rain gauge** to collect and measure the amount of rain that falls. These elements, along with the kinds of clouds in the sky, can help you predict the weather.

When you think of clouds, what kinds of clouds do you imagine? Different clouds come with different kinds of weather. Look below. Which kind of cloud brings rainy weather?

Guided Questions

What are four elements of weather that can be measured?

Cumulus
Fair-weather clouds that occur on sunny days

Stratus
Thick, dark clouds that produce rainy days

Cirrus
High, wispy clouds made of ice crystals mean fair but cold weather

Weather — Lesson 27

The Sun makes water evaporate from the surface of Earth. *Evaporate* means turning from a liquid to a gas. The liquid water disappears and turns into a gas called water vapor that goes into the air. This makes the air humid. It gets wetter. When humid air cools, water vapor turns back to liquid water and becomes tiny drops of water. A cloud forms from the water drops.

Clouds are billions of tiny droplets of water. A cloud that touches the Earth's surface is called **fog**. If the temperature is cold enough, clouds and fog might also have tiny bits of ice in them.

As the tiny water drops move around in a cloud, they bump into each other. Then they form bigger drops of water. Finally these big drops fall to Earth as rain. If the temperature is cold enough, the drops might become snow or sleet.

Not all clouds cause rain, sleet, or snow. Cirrus clouds are made of tiny pieces of ice. They stay higher in the sky than other clouds. They form only when the weather is cool and dry. Cumulus clouds are white, puffy clouds that form on warm sunny days.

Guided Questions

How does a cloud form?

What is inside **fog**?

DIRECTIONS For each question, write your answer in the spaces provided.

1. What are some tools that people use to predict weather?

   _____

   _____

2. How can you predict weather by looking at the sky?

   _____

   _____

3. What happens when the Sun shines on water?

   _____

   _____

Lesson 27 — Weather

Apply the NYS Learning Standards

DIRECTIONS Read the text below and answer the questions.

You know that raindrops reach the ground when many water droplets stick together and become heavy. Sometimes water vapor is very high in the sky. The air is cold up there. The water vapor forms into ice crystals instead of water droplets. It comes to Earth as snow or sleet.

Sometimes, raindrops fall toward the Earth, but they do not reach the ground. Instead, strong winds lift them high up into the air again. They freeze in the cold air and turn into tiny balls of ice. They start to fall through warm air, and they collect more water. Then wind blows them back up, and that water turns to ice.

The small ice balls may fall down and fly back up many times. They get larger with every movement up and down. Finally, they are so heavy that the wind cannot carry them. They fall to the ground as hail.

Hail often falls during the summer. It may seem strange to see balls of ice on the ground in hot weather. Those hailstones were formed high up in the sky.

1. What are three types of frozen water from the sky?

   _____

2. Why can hail form in the summer?

   _____
   _____

3. If hail moves up to cold air and down through warm air many times, what happens to it?

   _____

4. What is the difference between sleet and hail?

   _____
   _____
   _____

DIRECTIONS Choose the best answer for each question. Then circle the letter of the answer you have chosen.

1. What is weather?
   A the direction of the wind
   B the difference between Sun and rain
   C the temperature on Earth
   D the condition of the air outside

2. Which tool is used to measure wind speed?
   A wind vane
   B rain gauge
   C anemometer
   D thermometer

3. Which one does not help you predict the weather?
   A temperature
   B evaporation
   C direction of wind
   D types of clouds

4. What causes water vapor to turn into tiny drops of water?
   A sunlight
   B humid air
   C evaporation
   D cool temperatures

5. Arrow 1 on this diagram shows
   A liquid water turning to water vapor
   B water vapor turning to liquid water
   C cumulus clouds turning into cirrus clouds
   D cirrus clouds turning into stratus clouds

Focus on the NYS Learning Standards

Lesson 28 Water Cycle

PS2.1c Water is recycled by natural processes on Earth.
- evaporation: changing of water (liquid) into water vapor (gas)
- condensation: changing of water vapor (gas) into water (liquid)
- precipitation: rain, sleet, snow, hail
- runoff: water flowing on Earth's surface
- groundwater: water that moves downward into the ground

LE6.2c Heat energy from the Sun powers the water cycle.

You can learn how Earth's water moves from land to air and back to land.

The **water cycle** is the continuous movement of water from land and bodies of water to air and back again.

Evaporation is what happens when a liquid heats up and turns it into a gas.

Condensation is what happens when water changes from a gas into a liquid. When water vapor gets cold, it forms water drops in clouds.

Precipitation is water that falls to Earth as rain, hail, sleet, or snow.

Groundwater is water from precipitation that soaks into the ground.

Guided Instruction

DIRECTIONS Read the following information.

Have you ever wondered how old the water you drank yesterday was? If you guessed that it was very old, you were correct. The water in every glass you drink has been around almost as long as Earth has, and Earth is billions of years old!

Earth has a limited amount of water. This means that the amount of water on Earth stays the same throughout time. Water is not added or taken away. It gets cycled over and over in a process called the water cycle. In the **water cycle**, water moves from land and bodies of water to air and back to land again.

Energy from the Sun powers the water cycle. Heat from sunlight warms the water in oceans, lakes, rivers, and ponds. As water heats up, it evaporates and turns from a liquid to an invisible gas called water vapor. This process is called **evaporation**.

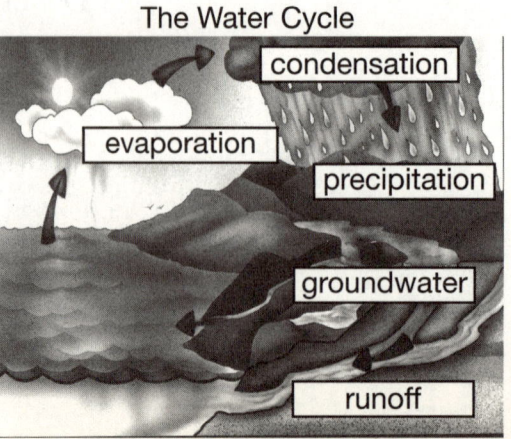

The Water Cycle

Guided Questions

How old is the water you drink?

What is the **water cycle**?

What happens during the process of **evaporation**?

Water Cycle — Lesson 28

The next step in the water cycle involves a change in water vapor. High in the sky, air temperatures get very cool. As water vapor in the air cools, it changes from a gas back into a liquid in a process known as **condensation**. During condensation, tiny water drops form in the air. The water drops come together to form clouds.

Eventually so much water vapor condenses that the water drops become large. When they become too large to stay in the air, they fall back to Earth as **precipitation**. Precipitation may be in the form of rain, hail, sleet, or snow.

When water falls back to Earth as precipitation, some of it returns directly to oceans, lakes, rivers, and ponds as runoff. Some of the water, however, ends up on land. It soaks into the soil and becomes **groundwater**. Plants take in groundwater through their roots. They release water back into the air through their leaves.

Guided Questions

How does water vapor form water drops?

What are different forms of **precipitation**?

What is **groundwater**?

DIRECTIONS For each question, write your answer in the spaces provided.

1. Through which process does water move from the ocean to the air?

2. What provides the energy for the water cycle?

3. What happens when water drops become too large?

4. What happens to precipitation?

Lesson 28 — Water Cycle

Apply the NYS Learning Standards

DIRECTIONS Read the paragraph, study the table, and answer the questions.

People depend on water for many different reasons. We use water in our homes to take showers, to clean laundry, and to brush our teeth. Restaurants use water to keep food and dishes clean. Farmers use water for crops. Earth has a limited amount of water. Therefore, it must be continuously recycled. However, sometimes water is used faster than it can be returned to the water cycle. Because of this, water conservation is important. To conserve water means to use only the water that you need and not waste it.

The table shows the amount of water used for certain activities in the home.

Activity	Amount of Water Used (gallons)
Taking one bath	25
Flushing the toilet once	5
Doing a load of laundry	30
A faucet leaking for one day	20

1. Which two activities use the most water?

2. Taking a bath uses twenty-five gallons of water. What else could you do to get clean and use less water?

3. Why is it important to fix all leaky faucets?

4. What are some things you can do at home to conserve water?

Water Cycle — Lesson 28

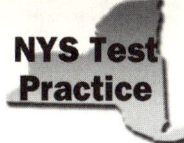

DIRECTIONS Choose the best answer for each question. Then circle the letter of the answer you have chosen.

1. By which process are clouds formed?
   A heating
   B evaporation
   C precipitation
   D condensation

2. The water cycle is powered by energy from
   A groundwater
   B the Sun
   C recycled water
   D the clouds

3. During hot, dry summer months, the amount of groundwater in an area may get very low. When this happens, which of the following most likely occurs?
   A Plants die.
   B The ground floods.
   C More rain falls.
   D The water cycle stops.

4. Which process is shown by the arrow in the picture?
   A precipitation
   B groundwater
   C evaporation
   D runoff

5. Which activity listed below is least important and could be skipped to conserve water?
   A cleaning dishes
   B washing a car
   C doing laundry
   D taking a shower

Focus on the NYS Learning Standards

Lesson 29 Erosion

PS2.1d Erosion and deposition result from the interaction among air, water, and land.
- interaction between air and water breaks down earth materials
- pieces of earth material may be moved by air, water, wind, and gravity
- pieces of earth material will settle or deposit on land or in the water in different places
- soil is composed of broken-down pieces of living and nonliving earth material

You can learn how Earth's surface changes over time.

Weathering is a process that breaks down rocks and wears down Earth's surface.

Erosion occurs when natural forces, such as wind and water, move Earth's materials.

Glaciers are sheets of ice that move slowly, causing erosion to the surrounding land.

Guided Instruction

DIRECTIONS Read the following information.

Earth's surface is constantly changing. Natural forces wear away at Earth's surface over many years. Most of these changes do not occur quickly.

Changes to Earth's surface happen when rocks are broken down into smaller pieces. This process is called **weathering**. Rocks may be broken down, or weathered, by wind, water, ice, or plant action.

Strong, heavy winds carry sand and other small pieces of rock. These small particles of rock grind away at larger rocks. Abrasion is the wearing away of rock by the scraping or rubbing of rock against rock. This slowly breaks down the rocks into small pieces. The moving ice in **glaciers** also causes rocks to break apart.

Most rocks have cracks or small spaces in them. Rainwater can fall or run into the cracks and freeze. As the water freezes and turns to ice, it expands and breaks up the rock.

Tree growth can also cause weathering. Tree roots can make their way into the spaces between rocks. As the roots grow, they push on the rocks. This eventually breaks the rocks apart.

Guided Questions

What are four things that can cause **weathering**?

How do strong winds cause **weathering**?

Erosion **Lesson 29**

Chemicals that combine with rainwater also can break down rocks into smaller pieces. As the rainwater flows over rocks, the chemicals dissolve or wear them away. Weathered rocks move from one place to another in a process called **erosion**. Moving water in rivers or streams is the main cause of erosion. Wind and glaciers also cause erosion. Glaciers move slowly, pushing rocks ahead of them. Sometimes gravity causes erosion.

Guided Questions

What is the main cause of **erosion**?

Weathering and erosion work together. They continuously change Earth's surface by wearing down and carrying away rocks. Given enough time, weathering and erosion can wear away even the largest mountain.

DIRECTIONS For each question, write your answer in the spaces provided.

1. How does weathering change Earth's surface?

   _____

   _____

2. Describe the way in which freezing water can cause rocks to break apart.

   _____

   _____

3. What is the difference between weathering and erosion?

   _____

   _____

4. What do you think happens to small pieces of rock after rivers carry them away?

   _____

   _____

Lesson 29: Erosion

Apply the NYS Learning Standards

DIRECTIONS Read the paragraph, study the table, and answer the questions.

Weathering and erosion can create unique land features. Some of these land features are described in the table below.

Land Feature	Description	How It May Be Formed
V-shaped valley	Large, V-shaped area carved out of the land	As a river moves across the land, it slowly carves out a valley.
Delta	Area of land near the end of a river along the shore	Where a river flows into an ocean, it slows down, and the rocks and mud are deposited along its shore.
Mesa	Flat-topped mountain surrounded by steep cliffs	A strong rock layer protects rocks under it from erosion. As a result, water erodes away the weaker rocks around it, creating steep sides and leaving the strong rock layer as a flat top.

1. Barry went on a field trip with his class to the Grand Canyon. He drew his observations of the famous landform in his notebook. Use the table to describe what type of landform the Grand Canyon is and how it formed.

   _____

2. You are on a hike when you observe a mountain with steep cliffs and a flat top. Describe in your own words how this structure formed.

   _____
   _____

3. Explain how erosion helps build a delta.

   _____
   _____

4. Kristi's class wants to research V-shaped valleys. They have the following materials: a baking pan, sand, and a hose. How can they set up an experiment for their research?

   _____
   _____

Erosion **Lesson 29**

DIRECTIONS Choose the best answer for each question. Then circle the letter of the answer you have chosen.

1 Which best describes weathering?
  A freezing of water
  B wearing away of rocks and soil
  C moving rocks and soil
  D hot liquid rock coming up to Earth's surface

2 What is the cause of most land erosion?
  A moving water and wind
  B animals
  C trees
  D moving ice

3 Chemicals mixed with rainwater can change rocks through which action?
  A moving
  B smashing
  C dissolving
  D grinding

 4 If the riverbed area shown below receives heavy rains for several years in a row, which will most likely happen?

  A weathering only
  B both weathering and erosion
  C erosion only
  D a glacier will move

 5 Which land feature does the arrow point to?

  A mesa
  B delta
  C valley
  D volcano

Focus on the NYS Learning Standards

Lesson 30 Extreme Natural Events

PS2.1e Extreme natural events (floods, fires, earthquakes, volcanic eruptions, hurricanes, tornadoes, and other severe storms) may have positive or negative impacts on living things.

You can understand how natural events affect living things.

The **crust** is Earth's outer layer of solid rock.

The **mantle** is the thick layer of hot rock just beneath Earth's crust.

An **earthquake** happens when large rocks break and cause the Earth to shake.

A **volcanic eruption** happens when magma bursts through a crack in the Earth.

A **hurricane** is a strong storm with powerful winds.

A **flood** happens when water flows over the boundaries of a body of water.

A **tornado** is a strong, fast-turning column of air.

A **forest fire** is an uncontrolled fire in a wooded area.

Guided Instruction

DIRECTIONS Read the following information.

Suppose you slice Earth in half. You would see that the outer layer is thin and made of rock. This layer is called the **crust**. The crust sits on the **mantle**, a layer of hot rock. The mantle is so hot that a little bit of it is liquid. This melted rock is called magma.

Earth's crust and the top part of the mantle are like a jigsaw puzzle. They are made of large pieces of rock that fit together. These pieces slide past or under each other very slowly.

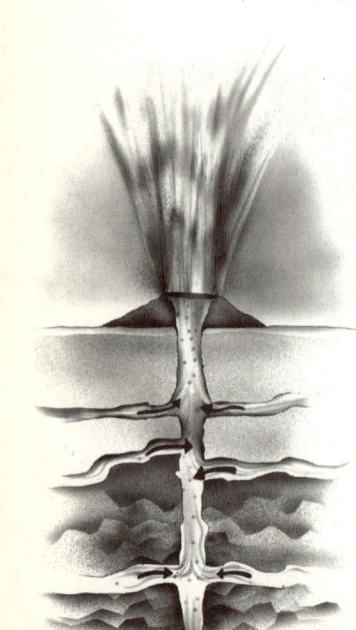

Sometimes the pieces get stuck. They have to move very suddenly to break free. This sudden movement causes an **earthquake**. When an earthquake happens, Earth shakes. Large earthquakes change Earth's surface. They can cause cracks and destroy buildings and roads.

The hot magma is under pressure below the crust. Sometimes it bursts through a crack in the crust. We call this a **volcanic eruption**. On top of the crust, the magma flows as lava. It can burn homes and trees.

Guided Questions

What is the Earth's **mantle** made of?

What is an **earthquake**?

What can lava do to trees?

Extreme Natural Events — Lesson 30

Forces from the Earth can cause changes. So can forces from the air. There are many kinds of storms that cause changes. A **hurricane** is a strong storm. Hurricanes arise over warm bodies of water. They produce rain and very strong winds. The winds and rain can cause harm to living things. A hurricane can cause a **flood** that destroys beaches and fills buildings with water.

Guided Questions

What is the difference between a **hurricane** and a **tornado**?

Thunderclouds are dark clouds that are filled with rain. Sometimes a tornado drops down from a cloud. A **tornado** is a whirling wind. It moves at a very fast speed. When a big tornado touches the ground, it can pull up trees and knock roofs off of houses.

The lightning from a storm can set off a **forest fire**. This is a wildfire that burns brush and trees in a wooded area. The wind can make a forest fire grow and spread. Plants and animals will die. Some animals will need to find new homes.

Lesson 30 Extreme Natural Events

The forces of nature can be harmful. They can have negative results. Some can also be helpful. They can have positive results. Lava from a volcanic eruption can build up the land. Some volcanic eruptions form islands in the ocean. A flood may deposit rich soil along a riverbank. This allows plants to grow. It makes rich farmland along rivers and streams. A forest fire can clear brush away. That helps young trees to get sunlight and grow taller. Some trees even depend on the hot temperatures of forest fires to release their seeds.

Guided Questions

How can a forest fire be harmful? How can it be helpful?

DIRECTIONS For each question, write your answer in the spaces provided.

1. What is magma?
   _____
   _____

2. What causes an earthquake?
   _____
   _____

3. What causes a volcanic eruption?
   _____
   _____

4. How can floods be helpful for farmers?
   _____
   _____

Extreme Natural Events | **Lesson 30**

Apply the NYS Learning Standards

DIRECTIONS Read the text below and answer the questions.

The San Francisco earthquake of 1906 was very strong. The land along the ocean moved to the north. The land further inland moved to the south. The land shifted nearly 20 feet. The shift was nearly 300 miles long.

Scientists learned a lot about earthquakes from the one in San Francisco. Many people who lived through the earthquake told about it. They wrote letters and articles. One man ran a music store. He found that pianos had been thrown around and broken. His set of breakable records had not been touched. One doctor wrote about the injured people he treated. Most came from an area of poorly-built buildings. The buildings were on land that had been built up by hand.

Land that was covered with dirt was badly damaged. Land that was mostly rock was less damaged. Land near the center of the shift was badly damaged. Land near either end was less damaged.

Scientists studied the earthquake. They learned which buildings survived the shaking. They came up with new ways to build stronger buildings. Today, all buildings in San Francisco must be built a certain way to make sure that the building can withstand an earthquake.

1. What did the land do during the San Francisco earthquake?

_____

2. Where did many of the injured people live?

_____

_____

3. What kind of land received the most damage?

_____

_____

4. Why must buildings in San Francisco be built a certain way?

_____

_____

Lesson 30 Extreme Natural Events

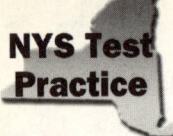

NYS Test Practice

DIRECTIONS Choose the best answer for each question. Then circle the letter of the answer you have chosen.

1. A forest fire might be helpful because there will be
   A no more plant seeds
   B damage to buildings
   C animals having to move
   D more sunlight for new plants

2. Which best describes a tornado?
   A strong winds hitting the seashore
   B a whirling wind connected to a cloud
   C a storm with heavy rain or snow
   D hot magma bursting through a crack

3. One result of a hurricane might be
   A an earthquake
   B a forest fire
   C a flood
   D a tornado

4. In the San Francisco earthquake of 1906,
   A magma pushed through cracks in the Earth
   B sections of Earth shifted in different directions
   C strong winds arose over warm water
   D whirling winds destroyed poorly-made houses

5. The island in the picture might be a good result from
   A a tornado
   B a flood
   C a volcanic eruption
   D a forest fire

Focus on the NYS Learning Standards

Lesson 31: Earth, the Moon, and the Sun

PS1.1a Natural cycles and patterns include: Earth spinning around once every 24 hours (rotation), resulting in day and night; Earth moving in a path around the Sun (revolution), resulting in one Earth year; the length of daylight and darkness varying with the seasons; weather changing from day to day and through the seasons; the appearance of the Moon changing as it moves in a path around Earth to complete a single cycle.

PS1.1b Humans organize time into units based on natural motions of Earth: second, minute, hour, week, month.

PS1.1c The Sun and other stars appear to move in a recognizable pattern both daily and seasonally.

You can learn about the way that Earth and other objects in space move.

Revolution is the movement of one object around another.

Rotation is the turning or spinning of an object around a central line, or axis.

An **axis** is an imaginary central line around which an object rotates.

The **Northern Hemisphere** of Earth is the area north of the equator.

The **Southern Hemisphere** of Earth is the area south of the equator.

The **equator** is an imaginary circle around Earth halfway between the North Pole and South Pole.

The Earth's Moon appears to have different shapes, called **phases**.

A **lunar cycle** is the period of time it takes the Moon to revolve around Earth once.

Guided Instruction

DIRECTIONS Read the following information.

Objects in space are in constant motion. Even though you cannot feel it, Earth is moving at this very moment. Earth moves around the Sun, and it spins around like a top.

When an object moves around another object, it revolves around it. The Sun is at the center of our solar system. Earth and other planets revolve around the Sun. The Moon revolves around Earth. This type of movement is called **revolution**. Objects take different amounts of time to complete a revolution.

Guided Questions

What does it mean when an object completes a **revolution**?

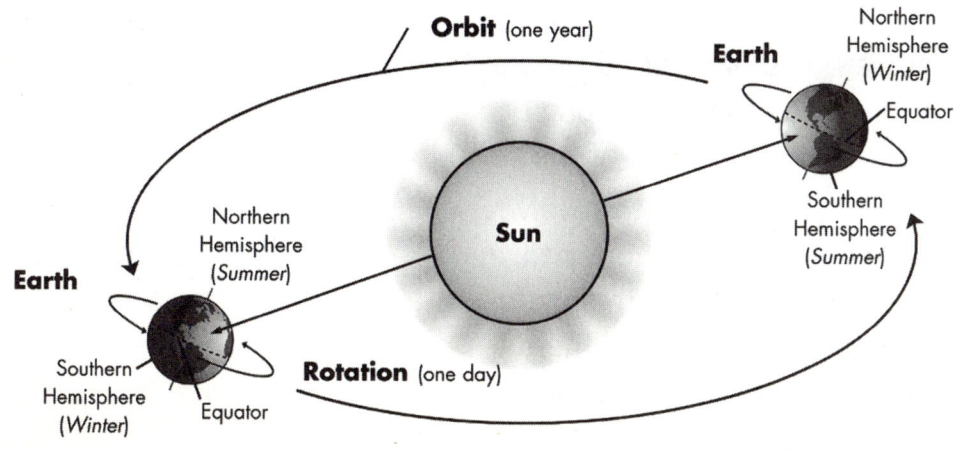

Lesson 31 Earth, the Moon, and the Sun

It takes Earth about 365 days, or one year, to make one revolution around the Sun. The farther a planet is from the Sun, the longer it takes to complete a revolution. One *planetary year* is defined as the amount of time it takes a planet to revolve around the Sun.

Objects in space also move by spinning. This type of movement is called **rotation**. Revolution and rotation take place at the same time. As planets revolve around the Sun, they also spin, or rotate. The Moon rotates as it revolves around Earth.

When an object rotates, it turns around a central line called an **axis**. Each planet takes a certain amount of time to rotate once around its axis. One planetary day is the amount of time it takes the planet to rotate around its axis. Earth completes one rotation in about twenty-four hours. Venus, however, takes 243 hours to complete one rotation.

Earth is divided into two sections. They are called the **Northern Hemisphere** and the **Southern Hemisphere**. These two sections are separated by an imaginary circle around Earth, called the **equator**. It is halfway between the North Pole and South Pole. The United States is located in the Northern Hemisphere.

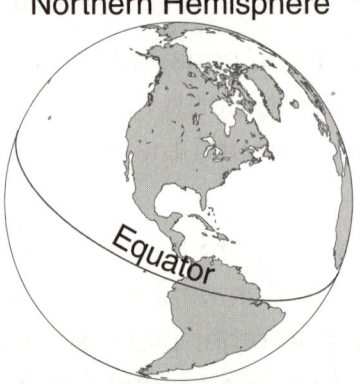

Earth rotates on its axis as it revolves around the Sun. However, Earth's axis does not run in a straight up-and-down line. It is tilted, or turned slightly on its side. Because of this tilt, some parts of Earth get more sunlight than other parts at any one time. The tilting of Earth's axis is what gives us the four seasons of the year.

It takes Earth twelve months, or one year, to revolve around the Sun. During the six months that the North Pole is tilted toward the Sun, the Northern Hemisphere gets more sunlight and longer days than the Southern Hemisphere. People living in the Northern Hemisphere

Guided Questions

How long does it take Earth to **revolve** around the Sun?

What kind of motion is **rotation**?

What is an **axis**?

Where is the **equator**?

Earth, the Moon, and the Sun — **Lesson 31**

experience the warmer seasons of late spring, summer, and early fall during this time. People living in the Southern Hemisphere, however, experience the colder seasons of late fall, winter, and early spring at the same time. The Northern Hemisphere experiences these colder seasons in the six months that the North Pole is tilted away from the Sun.

It takes about twenty-eight days, or about one month, for the Moon to revolve around Earth. It takes the same amount of time for the Moon to rotate around its own axis. Because these time periods are the same, we always see the same side of the Moon from Earth. However, as the Moon revolves around Earth, we often see different portions of the Moon's lighted part.

Sometimes the Moon looks like a circle. Other times it looks like a half circle, a crescent, or something in between. These different shapes of the Moon are called **phases**. The phases repeat about every twenty-eight days, or after one complete revolution around Earth. This period of time is called a **lunar cycle**.

Guided Questions

Which hemisphere gets more sunlight when the North Pole is tilted toward the Sun?

How long is a **lunar cycle**?

DIRECTIONS For each question, write your answer in the spaces provided.

1. Describe the two kinds of movements made by planets.

   _____

   _____

2. How long does it take Earth to make one revolution around the Sun?

   _____

3. Which seasons take place in the Northern Hemisphere when the North Pole is tilted away from the Sun?

   _____

Lesson 31 Earth, the Moon, and the Sun

Apply the NYS Learning Standards

DIRECTIONS Read the paragraph, study the chart, and answer the questions.

The four closest planets to the Sun are called the inner planets. They are Mercury, Venus, Earth, and Mars. The chart below shows the length of one planetary day and one planetary year for each inner planet.

Planet	Approximate Length of One Planetary Day	Approximate Length of One Planetary Year
Mercury	58.6 hours	88 days
Venus	243 hours	225 days
Earth	24 hours	365 days
Mars	24.5 hours	687 days

1. What is a planetary day?

   _____

   _____

2. How long does it take Mercury to complete one revolution around the Sun?

   _____

   _____

3. Which inner planets take longer than Earth to complete one rotation?

   _____

   _____

4. Neptune is the farthest planet from the Sun. Would you expect its planetary year to be longer or shorter than Earth's? Explain.

   _____

   _____

   _____

Earth, the Moon, and the Sun **Lesson 31**

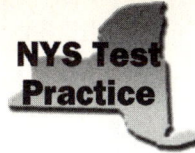

DIRECTIONS Choose the best answer for each question. Then circle the letter of the answer you have chosen.

1 Which of the following do planets in our solar system revolve around?

A Earth

B the Sun

C Mars

D the Moon

2 About how long does it take the Moon to rotate once on its axis?

A one day

B one week

C one month

D one year

3 Which of the following causes the seasons?

A the Sun's rotation

B the tilt of the Earth's axis

C the rising and setting of the Sun

D the phases of the Moon

 4 Which of the following is equal to one planetary day?

A one revolution

B one axis

C one rotation

D one orbit

5 How many hours does it take Earth to complete one rotation?

A 12

B 24

C 36

D 48

 6 Suppose daylight lasted for thirteen hours on June 21. How many hours of nighttime would occur on that day?

A 11

B 12

C 24

D 36

Performance Task

Make Your Own Compass

Focus on the NYS Learning Standards: PS5.1a, e, PS5.2a, b, T1.4a, b, T1.5a, c, ICT6.2, ICT 6.4, IPS7.1, IPS7.2

Task:

Build a simple compass.

Earth is a giant magnet. One of its magnetic poles is near the North Pole. The other is near the South Pole. You can use Earth's magnetic poles to figure out which direction is north. To do this, you need a tool called a compass. A compass has a magnet in it.

Materials:

- magnet with poles labeled "N" and "S"
- steel needle
- small foam ball
- dish of water
- compass

Directions:

1. Carefully pick up the needle because one end is sharp. Rub one pole of the magnet along the needle in the same direction 20 times. This should magnetize the needle, or turn it into a magnet.

2. Place the needle on your desk. Take the "S" end of the magnet and place it near the needle. Make sure the magnet doesn't touch the needle. One end of the needle should swing around toward the magnet. This shows that the needle is magnetized. If this doesn't happen, repeat Step 1. If it does happen, go to Step 3.

3. Which end of the needle points toward the "S" end of the magnet? In the drawing below, label that end "N."

4. Ask your teacher to push the needle through the foam ball.

5. Place the dish of water in front of you on your desk. Then place the foam ball in the dish of water. The ball should float in the water. The needle should point across the water, as the picture below shows. If the needle does not point across the water, ask your teacher to help you fix the needle.

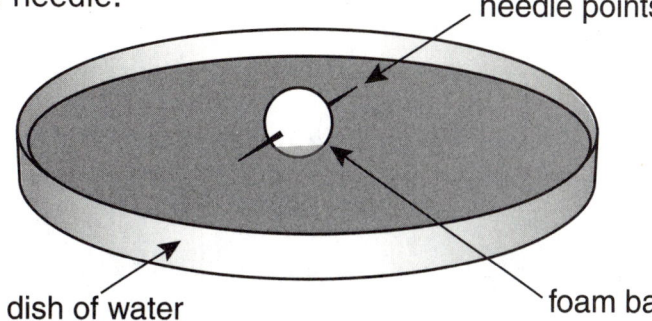

6. Take your hands away from the dish. Let the ball and needle come to a rest. In the space below, make a drawing that shows which way the needle is pointing. In your drawing, be sure to show your desk, the bowl of water, the foam ball, and where you are sitting. Label the "N" end of the needle.

7. The "N" end of the compass is pointing toward Earth's North Pole. What does this tell you about the North Pole?

_____

_____

8. Scientists think that Earth's magnetic field comes from its very center part, called the core. What do you think Earth's core is made of?

_____

9. When would it be useful to have a compass?

_____

_____

10. Use your compass to try to figure out which direction in your classroom is north. Then place a magnet near your compass. What happens? Explain why you think this happens.

_____

_____

11. Why do you think it is important to stay away from other magnetized objects when you are trying to use a compass?

_____

_____

_____

12. If Earth is a magnet, why do you think all metal objects do not move toward the North or South Poles?

_____

_____

13. Some planets, Earth for example, are giant magnets. Other planets are not.

Planet	Is it a Giant Magnet?
Mercury	yes, but it is very weak
Venus	no
Earth	yes
Mars	no, but it might have been one long ago
Jupiter	yes, and it is the strongest in the Solar System
Saturn	yes
Uranus	yes
Neptune	yes

14. Look at the table above. On which planets would a compass not work?

_____

_____

Building Stamina

Part I

Directions (1–16): Each question is followed by four choices. Decide which choice is the best answer. Circle the letter of the answer you have chosen.

1. By which process does water from clouds fall to Earth?
   A precipitation
   B cooling
   C condensation
   D evaporation

2. Which of the following causes rock to weather by dissolving it?
   A other rocks
   B chemicals
   C earthworms
   D ocean waves

3. How does abrasion cause rocks to break down?
   A Rocks grind away at other rocks.
   B Water freezes and breaks rocks apart.
   C Rocks get buried and squeezed by sediment.
   D Acids dissolve rocks into smaller pieces.

4. Which change occurs during the process in which clouds are formed?
   A liquid to gas
   B solid to liquid
   C gas to liquid
   D liquid to solid

5. Water that falls to Earth as precipitation and soaks into the soil becomes
   A salt water
   B condensation
   C freezing rain
   D groundwater

6. In which type of day is there the most evaporation?
   A sunny
   B cold
   C wet
   D cloudy

7 What happens to water from a lake during the process of evaporation?

   A It becomes water vapor.
   B It soaks into the ground.
   C It enters the oceans.
   D It falls to Earth as rain.

8 Which causes Earth's surface to shake?

   A hurricane
   B tornado
   C earthquake
   D flood

9 What are glaciers?

   A snow that stays on the tops of mountains
   B sheets of ice that move slowly downhill
   C ice that forms on the surface of oceans
   D currents of water that flow from the poles

10 What does the picture above show?

    A an earthquake
    B Earth's layers
    C water cycle
    D rock cycle

11 Which of these does not cause weathering?

    A light
    B wind
    C water
    D earthquakes

12. What do you call the condition of the air outside?

    A wind
    B hail
    C weather
    D tornado

13. Which is not a tool used to measure or describe weather?

    A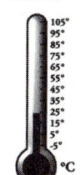

    B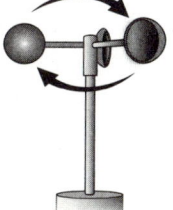

    C

    D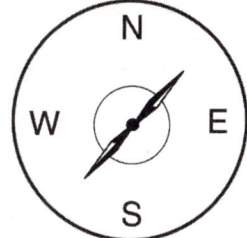

14. What do we call the period of time it takes the Earth to revolve around the Sun?

    A an hour
    B a day
    C a month
    D a year

15. The Sun appears to rise in the east because

    A the Earth revolves around the Sun
    B the Earth rotates around its axis
    C the Northern Hemisphere faces the Sun
    D the seasons change throughout the year

16. When the North Pole tips toward the Sun, which seasons do we have in the United States?

    A late spring, summer, early fall
    B late fall, winter, early spring
    C late winter, spring, early summer
    D late summer, fall, early winter

Part II

Directions (17–20): Record your answer on the lines provided below each question.

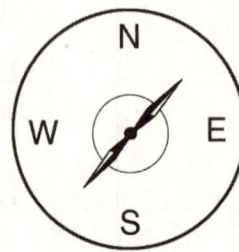

17 The needle on this compass is pointing toward the North Pole, but it is not pointing toward N on the compass. What would you need to do to make it point to N? Draw a picture and explain it in words.

_____

_____

18 How could you use this magnet to test whether the needle on the compass was truly pointing north?

_____

19 Why might the needle on the compass move away from north when you touch it with a magnet?

_____
_____

20 Unlike Earth, the planet Venus is not a giant magnet. How might your compass behave on Venus?

_____
_____

End-of-Book
Building Stamina®

The End-of-Book **Building Stamina**®
is a comprehensive review
of the New York State Learning Standards
covered in the lessons.
By practicing with these challenging,
broad-based, higher-level thinking questions,
you will build up your stamina to succeed
on the NYS Elementary-Level Science Test
and in other academic endeavors
that require higher-level thinking.

Building Stamina

Part I

Directions (1–29): Each question is followed by four choices. Decide which choice is the best answer. Circle the letter of the answer you have chosen.

1 Pam and Ron are working on a project about birds. Pam writes down the different types of birds she sees during recess. Ron does the same.

Ron's Data	Pam's Data
wood duck	blue jay
bald eagle	wood duck
red-bellied woodpecker	
blue jay	

Ron notices that Pam's data is different than his. What should he do?

A Write down only the birds he saw.

B Write down only the birds Pam saw.

C Write down only the birds he saw that Pam did not see.

D Write down only the birds Pam saw that he did not see.

2 Magnets are attracted to

A plastic

B rubber

C iron

D wood

3 Which month was the warmest?

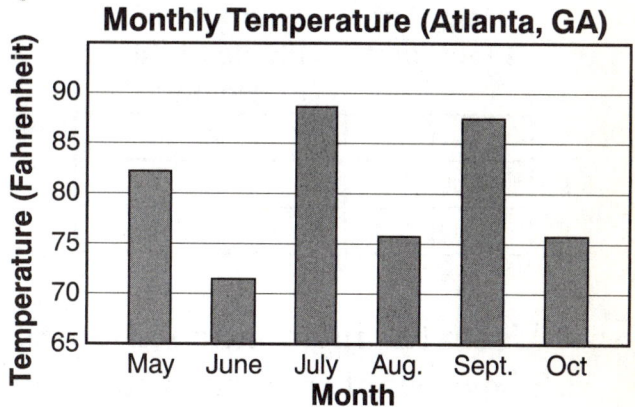

A May

B June

C July

D August

4 Squirrels live in trees in the forest. What will most likely happen to the squirrels if a forest fire burns down the trees?

A. They will have more babies.

B. They will find a new habitat.

C. They will need less food.

D. They will stay in the burned forest.

Mr. Powell's class observed 20 plants. They measured how tall each plant grew. The table shows their results.

Height of Plant (cm)	Number of Plants at this Height
0	5
2	5
6	10

5 Using the information in the table above, what fraction of the plants grew 2 centimeters?

A $\frac{1}{4}$

B $\frac{1}{3}$

C $\frac{2}{3}$

D $\frac{3}{4}$

6 Mark is doing a science activity. He decides to use a magnifying glass to help him. Mark is most likely studying something that

A is very big

B is very small

C happens very slowly

D happens very quickly

7 Nate is writing the procedure he used for a science project. Here is his notebook.

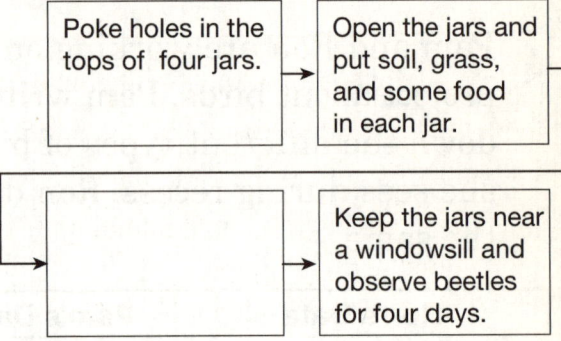

Which of these is most likely Step 3?

A Go outside and observe beetles.

B Push seeds down into the soil.

C Put one beetle in each jar.

D Write what happens in the jars every day.

8 Which of these describes how wind weathers rocks?

A Water flows over the rock.

B Wind blows sand against the rock.

C Minerals dissolving when it rains.

D Water evaporates on the rock.

Building Stamina

9 Sasha puts cold water in two jars. He puts insulation around Jar A. He does not put insulation around Jar B. He puts both jars in a warm room. What will most likely happen?

A The water in Jar A will get colder.

B The water in Jar B will get colder.

C The water in Jar A will stay cold longer than the water in Jar B.

D The water in Jar A will get warm more quickly than the water in Jar B.

10 What would happen if you opened the switch in the circuit shown below?

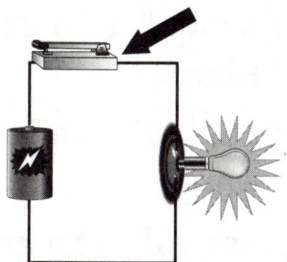

A The bulb would go out.

B The bulb would stay lit.

C The battery would run down.

D The bulb would go dim.

11 What change will happen when you boil water?

A The water will begin to change to a liquid.

B The water will begin to change to a solid.

C The water will begin to change to a mixture.

D The water will begin to change to a gas.

12 Which animal does not look like its mother when it is born?

A snake

B goldfish

C elephant

D caterpillar

13 What causes night and day on Earth?

A Earth's rotation on its axis

B Earth's revolution around the Sun

C changes in the phases of the Moon

D changes in the Sun as it orbits Earth

14. A student uses two balls of different sizes to model Earth and the Sun. The student uses a Ping-Pong ball as a model of Earth and a beach ball as a model of the Sun. What do these models show?

   A The Sun is the center of the solar system.
   B Earth is far away from the Sun.
   C The Sun is larger than Earth.
   D Earth orbits the Sun.

15. Which trait most helps a polar bear survive in a cold environment?

   A sharp teeth
   B a layer of fat
   C white fur
   D the ability to swim

16. Which is an example of a learned behavior?

   A A child has blue eyes.
   B A dog sits on command.
   C A raccoon hunts for food.
   D A plant grows toward light.

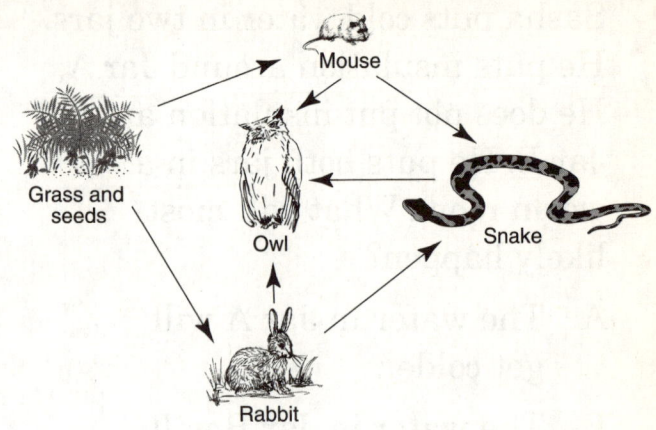

17. The arrows in the picture point from the food to the organism that eats it. Which two organisms are competing for the same food?

   A grass and rabbit
   B rabbit and mouse
   C owl and rabbit
   D mouse and owl

18. Which statement is an example of an observation?

   A The plant is brown because it needs water.
   B The plant will turn green if it is watered.
   C The plant has a flower.
   D The plant needs more sunlight.

19 The pictures above show different events in the life cycle of a tree. Which is the correct sequence of events?

A A, B, C, D
B D, B, C, A
C D, C, B, A
D C, D, B, A

20 The geese are flying to a warmer place to find food. This is an example of

A camouflage
B conservation
C hibernation
D migration

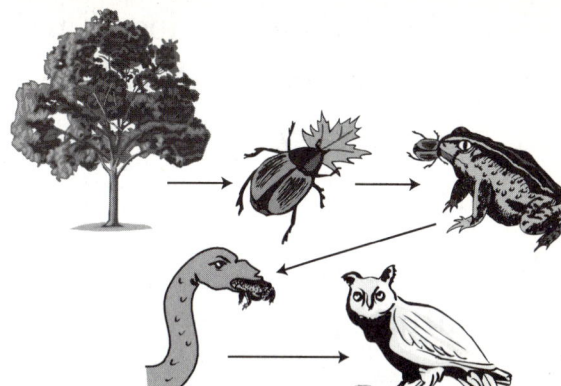

21 Which is missing from the food chain shown above?

A producer
B carnivore
C decomposer
D consumer

22 Two birdfeeders are each filled with a different kind of birdseed. A student wants to know which birdseed the birds like more. What should she measure?

A how many birds visit each feeder every hour
B how much birdseed is left in each feeder at the end of each day
C how long the birds stay at each feeder
D how many different types of birds visit each feeder

23 What kind of energy does Earth get directly from the Sun?

   A light energy
   B heat energy
   C electricity
   D natural gas

24 What is the main difference between a producer and a consumer?

   A A producer eats only other animals.
   B A producer makes its own food.
   C A consumer can live only on land.
   D Only a consumer is broken down by a decomposer.

25 Which of these properties cannot be easily observed?

   A color and shape
   B texture or feel
   C parts that make up a material
   D mass and length

26 Here is a picture of the water cycle. Which happens right before the precipitation falls to the ground?

   A Water evaporates from the lake.
   B Water vapor rises up in the air.
   C Water vapor condenses.
   D Water droplets join together.

27 The front entrance of a museum has some tall steps. Which of these could help a person in a wheelchair to go inside the museum?

   A inclined plane
   B pulley
   C lever
   D circuit

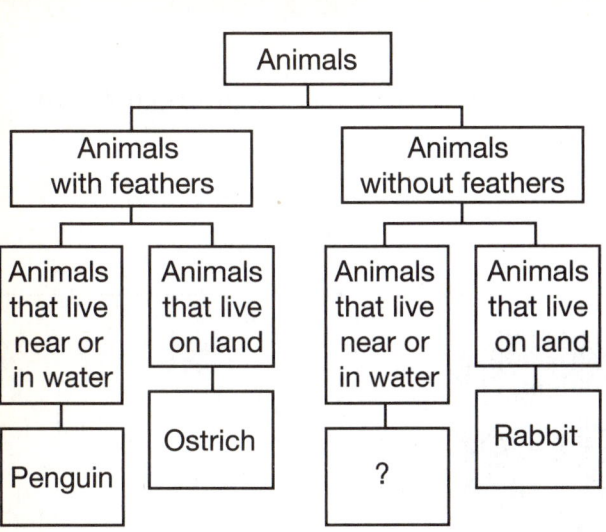

28 Which animal would fit in the empty box above?

A

B

C

D

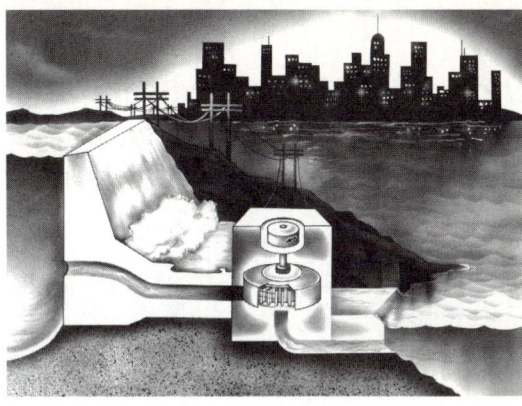

29 In the picture, the energy from moving water is being used to make

A electricity

B heat

C wind

D gas

Part II

Directions (30–47): Record your answer on the lines provided below each question.

30 What four things do plants need to live?

_____

31 Name one nonliving thing in this picture.

_____

32 Which plant above probably has longer roots? Why do you think so?

_____

33 Tell how the lifespan of an elephant is different from the lifespan of a squirrel.

_____

_____

Building Stamina

34 How do gills and fins help fish survive in their environment?

_____
_____

35 Where do plants get the energy they need to make food?

_____

36 What is one helpful effect of building a road? What is one harmful effect of building a road?

_____
_____
_____

37 Why does the Sun seem to rise in the east and set in the west?

_____
_____

38 What tool would you use to test the temperature of an object?

_____

39 The boy throws the ball up. Explain what causes it to come back down.

_____

40 Of your five senses, which four would you use to observe the properties of a potato?

_____

41 Solid and gas are two states of matter. What is the third?

_____

42 A magnet picks up a paper clip that is 1 cm away. It does not pick up a paper clip that is 6 cm away. Why not?

_____

_____

43 A scientist counted some bluebirds at a bird feeder. Later, she counted some goldfinches at the feeder. How can the scientist find how many birds she saw in all?

_____

_____

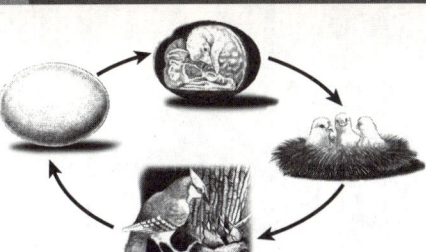

44 Explain these stages in a bird's life cycle. Tell about them in order.

_____
_____
_____

45 A student wants to test how tall sunflower seeds grow in three different kinds of soil. Name some materials the student will need for the test.

_____
_____
_____

46 Name one trait that you inherited from your parents.

_____
_____

47 What do the living things shown here have in common with all other living things?

_____

End-of-Book Performance Test

The End-of-Book Performance Test is a review of the New York State Learning Standards and Performance Tasks covered in the chapters. By practicing with these challenging Performance Tasks, you will build up your stamina to succeed on the NYS Elementary-Level Science Test and in other academic endeavors that require classroom experiments.

Performance Test

Station 1: Compare Rocks

Focus on the NYS Learning Standards: M3.1a, S1.1a, S2.3b, PS 3.1b, c, f

Task:

You will conduct an investigation in which you will study the differences among rocks that you find.

Materials:

- 4 rocks
- masking tape
- pen
- a piece of notebook paper
- metric ruler
- magnifying glass

Directions:

1. Collect four rocks that you would like to study. The rocks should be about the same size. Try to collect rocks that are different colors, shapes, and textures. You will describe their properties and compare them.

2. Use the masking tape and the pen to label each rock with a number from 1 to 4. This way, you will not get your rocks confused as you record your observations.

3. Create a table on your notebook paper. You will fill in this table as you make observations of your rocks. The table should look like the one below.

Rock	Color	Shape	Size	Texture
Rock 1				
Rock 2				
Rock 3				
Rock 4				

4. The first thing you see when you look at a rock is probably its color. Is it black? Is it gray? Is there more than one color? Record your observations in your table in the column labeled "Color."

5. Now, you will look at each rock's shape. Think about the shape of the rock. Is it round? Is it flat? Is it long? Record your observations in your table in the column labeled "Shape."

6. For this step, you will need your ruler. Measure each rock's length, width, and height in centimeters. Record your observations in the column labeled "Size."

7. Which rock is the longest?

   _____

8. Examine the texture of each rock. Does the rock feel rough or smooth? Use your magnifying glass to look at each rock. Do you see crystals? Are they big or small? Record your observations in the column labeled "Texture."

9. If you have a rock with big crystals, take a closer look at them with a magnifying glass. What color are they? What shape are they? How do the big crystals make the rock feel when you touch it?

   _____

   _____

10. Rub Rock 1 against Rock 2 to test hardness. Did Rock 1 scratch Rock 2? The rock that can scratch the other rock is harder. Do the same with Rock 3 and Rock 4. As you scratch each rock, complete the table below by circling Y or N for each pair of rocks to show what you learned.

Scratches →	Rock 1	Rock 2	Rock 3	Rock 4
Rock 1		Y N	Y N	Y N
Rock 2	Y N		Y N	Y N
Rock 3	Y N	Y N		Y N
Rock 4	Y N	Y N	Y N	

11. Which rocks are harder?

_____

_____

12. Explain how you figured out which rocks are harder.

_____

_____

Performance Test

Station 2: Test a Magnet's Power

Focus on the NYS Learning Standards: M1.2a, M2.1b, S1.3a, S2.3b, S3.2a, PS5.1e, PS5.2b

Task:

Test the power of a magnet.

A magnet has two poles, north and south. The entire magnet will attract metals with iron in them. Some parts of the magnet have a stronger attraction than others.

Materials:

- bar magnet with poles labeled "N" and "S"
- 20 paper clips
- inch ruler
- pen
- notebook paper

Directions:

1. Unfold one paper clip to make an "S" shape. This will be your paper clip hook.

2. Attach the paper clip hook to the north end of the magnet.

3. Carefully place paper clips on the paper clip hook. See how many paper clips the hook will hold before it slips off the magnet. Copy this table on your notebook paper. Record your findings on row 1 of the chart.

Position of Paper Clip Holder	Number of Paper Clips in All
North end of magnet	
1 inch from end of magnet	
Center of magnet	
1 inch from south end of magnet	
South end of magnet	

4. Use the ruler to measure 1 inch from the north end of the magnet. Move the paper clip hook to that spot. See how many paper clips it will hold before it slips off the magnet. Record your findings on the chart.

5. Repeat your test, placing the hook at the center of the magnet, 1 inch from the south end, and at the south end. Record your findings on the chart.

6. Which part of the magnet held the most paper clips?

    _____

7. Which part of the magnet held the fewest paper clips?

    _____

8. What conclusion can you draw about the power of a magnet? Where is most of its power?

    _____

    _____

9. Why do paper clips work well for this experiment?

_____

_____

10. A paper clip weighs about ½ gram. What is the greatest weight in grams that your magnet was able to hold? (Remember that the paper clip hook weighs ½ gram, too!)

_____

11. Use your findings from the chart. Make a bar graph like this one. Fill in the bars to show what you found.

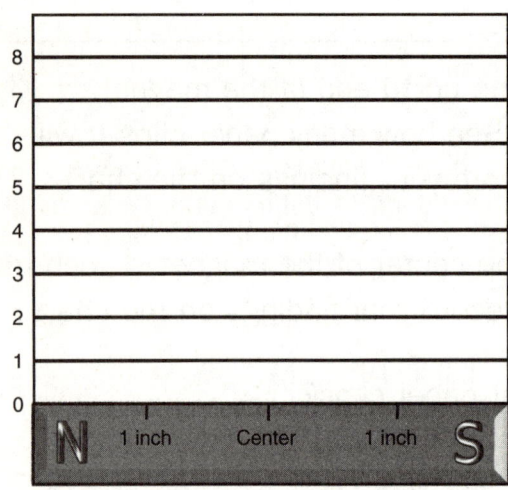

12. What shape does the bar graph show?

_____

_____

Performance Test

Station 3: Change Runoff to Groundwater

Focus on the NYS Learning Standards: S2.3a, PS2.1c, T1.4b, T1.5b, LE3.1b

Task:

Compare the effect of rainwater with and without plant roots in soil. Rainwater falls on the Earth. Some of it sinks into the ground to create groundwater. How do plants affect this?

Materials:

- 2 clear plastic cups
- masking tape
- marker
- inch ruler
- potting soil
- 3 craft sticks or popsicle sticks
- water
- measuring cup
- clock with second hand or egg timer

Directions:

1. Cut two 2-inch pieces of tape. Place one on each cup so that the top of the tape touches the top edge of the cup.

2. Mark each piece of tape ½ inch from the top of the cup.

3. Fill each cup with soil to the bottom edge of the tape. Pack the soil down.

4. Wiggle the three sticks into one cup's soil so that they nearly reach the bottom of the cup. Leave the other cup alone.

 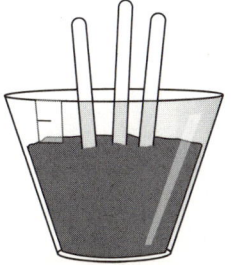

5. Fill the measuring cup with water. Pour water into each cup up to the mark on the tape.

6. After 1 minute, mark the level of the water on the tape.

7. Draw a picture of your results.

8. Which cup has a lower level of water?

   _____

9. Where did the water in that cup go?

   _____

   _____

10. As plants grow in soil, their roots make openings in the soil. In this experiment, what are you using to model the roots of plants?

    _____

11. As roots grow, what happens to rainwater that falls on Earth?

    _____

    _____

12. How might planting trees and bushes help to prevent flooding?

    _____

    _____

Properties of Common Minerals

LUSTER	HARD-NESS	CLEAVAGE	FRACTURE	COMMON COLORS	DISTINGUISHING CHARACTERISTICS	USE(S)	MINERAL NAME	COMPOSITION*
Metallic Luster	1–2	✔		silver to gray	black streak, greasy feel	pencil lead, lubricants	**Graphite**	C
Metallic Luster	2.5	✔		metallic silver	very dense (7.6 g/cm^3), gray-black streak	ore of lead	**Galena**	PbS
Metallic Luster	5.5–6.5		✔	black to silver	attracted by magnet, black streak	ore of iron	**Magnetite**	Fe$_3$O$_4$
Metallic Luster	6.5		✔	brassy yellow	green-black streak, cubic crystals	ore of sulfur	**Pyrite**	FeS$_2$
Either	1–6.5		✔	metallic silver or earthy red	red-brown streak	ore of iron	**Hematite**	Fe$_2$O$_3$
Nonmetallic Luster	1	✔		white to green	greasy feel	talcum powder, soapstone	**Talc**	Mg$_3$Si$_4$O$_{10}$(OH)$_2$
Nonmetallic Luster	2		✔	yellow to amber	easily melted, may smell	vulcanize rubber, sulfuric acid	**Sulfur**	S
Nonmetallic Luster	2	✔		white to pink or gray	easily scratched by fingernail	plaster of paris and drywall	**Gypsum** (Selenite)	CaSO$_4$·2H$_2$O
Nonmetallic Luster	2–2.5	✔		colorless to yellow	flexible in thin sheets	electrical insulator	**Muscovite Mica**	KAl$_3$Si$_3$O$_{10}$(OH)$_2$
Nonmetallic Luster	2.5	✔		colorless to white	cubic cleavage, salty taste	food additive, melts ice	**Halite**	NaCl
Nonmetallic Luster	2.5–3	✔		black to dark brown	flexible in thin sheets	electrical insulator	**Biotite Mica**	K(Mg,Fe)$_3$ AlSi$_3$O$_{10}$(OH)$_2$
Nonmetallic Luster	3	✔		colorless or variable	bubbles with acid	cement, polarizing prisms	**Calcite**	CaCO$_3$
Nonmetallic Luster	3.5	✔		colorless or variable	bubbles with acid when powdered	source of magnesium	**Dolomite**	CaMg(CO$_3$)$_2$
Nonmetallic Luster	4	✔		colorless or variable	cleaves in 4 directions	hydrofluoric acid	**Fluorite**	CaF$_2$
Nonmetallic Luster	5–6	✔		black to dark green	cleaves in 2 directions at 90°	mineral collections	**Pyroxene** (commonly Augite)	(Ca,Na)(Mg,Fe,Al)(Si,Al)$_2$O$_6$
Nonmetallic Luster	5.5	✔		black to dark green	cleaves at 56° and 124°	mineral collections	**Amphiboles** (commonly Hornblende)	CaNa(Mg,Fe)$_4$(Al,Fe,Ti)$_3$Si$_6$O$_{22}$(O,OH)$_2$
Nonmetallic Luster	6	✔		white to pink	cleaves in 2 directions at 90°	ceramics and glass	**Potassium Feldspar** (Orthoclase)	KAlSi$_3$O$_8$
Nonmetallic Luster	6	✔		white to gray	cleaves in 2 directions, striations visible	ceramics and glass	**Plagioclase Feldspar** (Na-Ca Feldspar)	(Na,Ca)AlSi$_3$O$_8$
Nonmetallic Luster	6.5		✔	green to gray or brown	commonly light green and granular	furnace bricks and jewelry	**Olivine**	(Fe,Mg)$_2$SiO$_4$
Nonmetallic Luster	7		✔	colorless or variable	glassy luster, may form hexagonal crystals	glass, jewelry, and electronics	**Quartz**	SiO$_2$
Nonmetallic Luster	7		✔	dark red to green	glassy luster, often seen as red grains in NYS metamorphic rocks	jewelry and abrasives	**Garnet** (commonly Almandine)	Fe$_3$Al$_2$Si$_3$O$_{12}$

*Chemical Symbols:
Al = aluminum   Cl = chlorine   H = hydrogen   Na = sodium   S = sulfur
C = carbon   F = fluorine   K = potassium   O = oxygen   Si = silicon
Ca = calcium   Fe = iron   Mg = magnesium   Pb = lead   Ti = titanium

✔ = dominant form of breakage

Commonly Used Units

LENGTH

Metric
1 kilometer = 1,000 meters
1 meter = 100 centimeters
1 centimeter = 10 millimeters

Customary
1 mile = 1,760 yards
1 mile = 5,280 feet
1 yard = 3 feet
1 foot = 12 inches

CAPACITY AND VOLUME

Metric
1 liter = 1,000 milliliters

Customary
1 gallon = 4 quarts
1 gallon = 128 ounces
1 quart = 2 pints
1 pint = 2 cups
1 cup = 8 ounces

MASS AND WEIGHT

Metric
1 kilogram = 1,000 grams
1 gram = 1,000 milligrams

Customary
1 ton = 2,000 pounds
1 pound = 16 ounces

TIME

1 year = 365 days
1 year = 12 months
1 year = 52 weeks
1 week = 7 days
1 day = 24 hours
1 hour = 60 minutes
1 minute = 60 seconds

Glossary

A

Adapt—to change to fit a certain environment (Lesson 26)

Adaptation—a trait that helps a living thing meet its needs or survive in its environment (Lessons 18, 21, 22)

Anemometer—a tool that measures wind speed (Lesson 27)

Attract—to pull toward something (Lesson 16)

Axis—an imaginary central line around which an object rotates (Lesson 31)

B

Balance—tool used to measure mass (Lesson 5)

Bar graph—graph that uses bars to show different amounts (Lesson 3)

C

camouflage—a body shape or color that helps an animal blend into its surroundings (Lesson 23)

Carnivores—consumers or animals who eat other animals (Lesson 24)

Circuit—the closed loop through which electricity travels (Lesson 10)

Classify—to group objects based on how they are alike or different (Lesson 7)

Community—a group of living things that work and live together (Lesson 26)

Compete—to try to get the same things from a habitat as something else (Lesson 23)

Conclusion—the answer you get to a question when you do an investigation (Lesson 1)

Condensation—what happens when water changes from a gas into a liquid (Lesson 28)

Consumers—living things that eat other living things for food (Lesson 24)

Crust—the Earth's outer layer of solid rock (Lesson 30)

Cultivate—to grow plants for a purpose (Lesson 26)

D

Data—facts gathered from an investigation (Lesson 3)

Decomposer—something that gets energy by breaking down decayed plants or animals (Lesson 24)

E

Earthquake—a shaking of the earth caused by large breaking rocks (Lesson 30)

Energy—the force that makes things move or change (Lessons 9, 10, 12)

Environment—all the living and nonliving things that surround an organism (Lessons 21, 26)

Equator—an imaginary circle around Earth halfway between the North Pole and South Pole (Lesson 31)

Erosion—the movement of Earth's materials by natural forces, such as wind and water (Lesson 29)

Evaporate—to change from liquid to gas (Lesson 11)

Evaporation—process when a liquid heats up and turns it into a gas (Lesson 28)

F

Fertilization—the first stage of a plant's growth (Lesson 19)

Flood—event when water flows over the boundaries of a body of water (Lesson 30)

Fog—a cloud that touches the Earth's surface (Lesson 27)

Food chain—the path of food from one living thing to the next (Lesson 25)

Force—a push or pull on an object that causes an object's position to change (Lessons 14, 15)

Forest fire—an uncontrolled fire in a wooded area (Lesson 30)

Friction—a force that stops an object from moving or slows the motion of a moving object (Lesson 14)

G

Gills—part of a fish that allows it to breathe underwater (Lesson 22)

Glaciers—sheets of ice that move slowly, causing erosion to the surrounding land (Lesson 29)

Gravity—a force that pulls objects down toward Earth's surface (Lesson 16)

Groundwater—water from precipitation that soaks into the ground (Lesson 28)

H

Hazard—something that is dangerous (Lesson 2)

Herbivores—consumers or animals that eat only plants (Lesson 24)

hibernate—what some animals do when they sleep through the winter (Lesson 23)

Hurricane—a strong storm with powerful winds (Lesson 30)

Hypothesis—your best guess to answer a science question (Lesson 1)

I

Inference—an explanation you make based on observations and reasoning (Lesson 1)

Inherited traits—body features or ways of acting that are received from parents (Lesson 18)

Insulation—a material that keeps heat from moving (Chapter 3 Performance Task)

L

Larva—the early, immature stage in the life cycle of some animals (Lesson 20)

Learned behaviors—traits that are not inherited from parents (Lesson 18)

Life cycle—the changes an organism goes through during life (Lesson 19)

Life span—the length of time from the beginning of a plant's development to its death (Lesson 19)

Line plot—graph that shows data along a number line (Lesson 3)

Living thing—something that grows, changes, and reproduces (Lesson 17)

Lunar cycle—the period of time it takes the Moon to revolve around Earth once (Lesson 31)

M

Magnet—object that attracts other objects that contain iron (Lesson 6)

Glossary

Magnetic—a characteristic which allows something to be pulled by a magnet from a distance (Lesson 16)

Magnetic poles—the opposite ends of a magnet (Lesson 16)

Mantle—the thick layer of hot rock just beneath Earth's crust (Lesson 30)

Mass—the amount of matter in an object (Lesson 6)

Math skills—choosing the correct operation to solve a problem (Lesson 4)

Matter—anything that has mass, takes up space, and has physical properties (Lessons 8, 9, 12)

Metamorphosis—a change in body form during the growth and development of some animals (Lesson 20)

Meterstick—a tool used to measure distance and length (Lesson 5)

Microscope—a tool used to observe and measure small objects (Lesson 5)

Motion—a change of position (Lesson 14)

N

Noise pollution—sound that is too loud or lasts too long and may be harmful (Lesson 13)

Nonliving thing—something that does not grow, change, or reproduce (Lesson 17)

Northern Hemisphere—area of Earth that is north of the equator (Lesson 31)

O

Observe—using your senses to notice and describe something (Lesson 1)

Omnivores—consumers or animals that eat both other animals and plants (Lesson 24)

Operations—processes in math that include addition, subtraction, multiplication, and division (Lesson 4)

P

Phases—different shapes of the Earth's moon (Lesson 31)

Physical property—a trait that can be observed with the senses (Lesson 8)

Picture graph—visual that uses symbols or pictures to show amounts (Lesson 3)

pollen—a powdery substance produced in flowers that helps plants reproduce (Lesson 21)

Position—the location of an object in relation to other objects (Lesson 14)

Precaution—something done to keep an accident from happening (Lesson 2)

Precipitation—water that falls to Earth as rain, hail, sleet, or snow (Lesson 28)

Predator—an animal that hunts and eats other animals (Lessons 22, 25)

Prediction—saying what you think will happen in an investigation (Lesson 1)

Prey—an animal that is hunted by other animals, or predators, for food (Lessons 22, 25)

Producers—living things that make their own food from sunlight, water, and air (Lesson 24)

Properties—traits you can observe about an object such as size, weight, color, and hardness (Lesson 7)

Pupa—a stage of metamorphosis when a larva changes into an adult (Lesson 20)

Glossary

R

Rain gauge—a tool that collects and measures amounts of rainfall (Lesson 27)

Repel—to push something away (Lesson 16)

Reproduce—to make more of the same kind of living thing (Lessons 17, 18)

Revolution—the movement of one object around another (Lesson 31)

Rotation—the turning or spinning of an object around a central line, or axis (Lesson 31)

S

Seed—a plant part that has a young plant and food stored inside (Lesson 19)

Seedling—a young plant that has sprouted (Lesson 19)

senses—sight, hearing, smell, taste, and touch (Lesson 6)

Shelter—a place that protects a living thing (Lesson 26)

Solar energy—energy from the Sun (Lesson 12)

Sound energy—mechanical energy that causes sound (Lesson 13)

Sound waves—vibrations in air or other matter, started by a vibrating object (Lesson 13)

Southern Hemisphere—area of Earth that is south of the equator (Lesson 31)

Species—one kind of living thing (Lesson 18)

Sprout—to begin to grow (Lesson 19)

States of matter—the forms of matter such as solid, liquid, and gas (Lesson 8)

Survive—to remain alive (Lesson 18)

T

Temperature—the measure of how cold or hot something is (Lesson 8, 11)

Thermometer—a tool used to measure temperature. (Lessons 5, 27)

Tornado—a strong, fast-turning column of air (Lesson 30)

Trait—a characteristic or feature that is passed down from one living thing to another (Lessons 17, 18, 22)

Transfer—the movement of energy from one place to another during which the energy may change form (Lesson 10)

Trial—a completed measurement (Lesson 5)

V

Vibrate—to move back and forth (Lesson 13)

Volcanic eruption—when magma bursts through a crack in the Earth (Lesson 30)

Volume—the amount of space matter takes up (Lesson 8)

W

Water cycle—the continuous movement of water from land and bodies of water to air and back (Lesson 28)

Weathering—a process that breaks down rocks and wears down Earth's surface (Lesson 29)

Wind vane—a tool that shows the direction of the wind (Lesson 27)

Notes